REWIRE

REWIRE

Digital Cosmopolitans in
the Age of Connection

ETHAN ZUCKERMAN

W. W. NORTON & COMPANY

NEW YORK LONDON

For Drew,
who will grow up in a world
as wide as his dreams.

Photograph credits: p. 57: Ethan Zuckerman; p. 62: Amy Balkin,
© Exploratorium, www.exploratorium.edu; p. 63: courtesy of Princeton Univer-
sity Library; p. 67: John O'Sullivan; p. 80: map by Ethan Zuckerman and Rahul
Bhargava; p. 212: Pierre Henry Chombart de Lauwe; p. 213: map rendered by Rahul
Bhargava from data from Zach Seward, map tiles by Stamen Design, map data by
OpenStreetMap.

For information about permission to reproduce selections from this book,
write to Permissions, W. W. Norton & Company, Inc.,
500 Fifth Avenue, New York, NY 10110

For information about special discounts for bulk purchases, please contact W. W.
Norton Special Sales at specialsales@wwnorton.com or 800-233-4830

Manufacturing by RR Donnelley, Harrisonburg
Book design by Lovedog Studio
Production manager: Julia Druskin

Library of Congress Cataloging-in-Publication Data

Zuckerman, Ethan.
Rewire : digital cosmopolitans in the age of connection / Ethan Zuckerman.
    pages cm
Includes bibliographical references and index.
ISBN 978-0-393-08283-8 (hardcover)
1. Social media. 2. Internet—Social aspects. 3. Cosmopolitanism. I. Title.
HM742.Z83 2013
302.23'1—dc23
                    2013007124

W. W. Norton & Company, Inc.
500 Fifth Avenue, New York, N.Y. 10110
www.wwnorton.com

W. W. Norton & Company Ltd.
Castle House, 75/76 Wells Street, London W1T 3QT

1 2 3 4 5 6 7 8 9 0

CONTENTS

It is hardly possible to overstate the value in the present state of human improvement of placing human beings in contact with other persons dissimilar to themselves, and with modes of thought and action unlike those with which they are familiar.

—*John Stuart Mill*

We are in great haste to construct a magnetic telegraph from Maine to Texas; but Maine and Texas, it may be, have nothing important to communicate.

—*Henry David Thoreau*

SECRETS AND MYSTERIES

THE SEVENTY-FIVE-YEAR-OLD AYATOLLAH RUHOLLAH KHO-meini had been exiled from Iran for fourteen years. His relentless critiques of Shah Reza Pahlavi, Iran's autocratic leader, had led to the ayatollah's expulsion, but had not silenced him. In 1977 he was living in neighboring Iraq, where he found a new way to share his message. Late in the evening, usually around ten, after the masses of pilgrims who'd come to visit the shrine of Imam Ali had left for the day, the ayatollah presented long lectures to anyone who would listen. The speeches were anti-shah diatribes, filled with conspiracy theories that tied the shah's Westernizing reforms to "the Jews and the Cross-worshipers" who sought to humiliate and subjugate Iran.

A few Iranians—no more than twelve hundred a month—were allowed to visit Iraq to worship at the shrine, and a small number of them returned home with an unusual souvenir: a cassette recording of the ayatollah's sermons. These cassettes were copied and freely distributed in the streets of Tehran and other Iranian cities. Pressured by President Jimmy Carter of the United States to live up to his promises of reform, the shah instructed his secret police, SAVAK, not to seize or destroy copies. The tapes were marked "Sokhanrani Mazhabi"—religious lecture—and sold next to tapes from the popular singers of the day. Parviz Sabeti, head of SAVAK's "antisubversion" unit estimated that more than 100,000 sermon cassettes were sold in 1978[1] and that millions of Iranians might have heard Khomeini's anti-shah invective.

Amir Taheri was editor of the pro-shah newspaper *Kahyan* when the tapes became popular. Two of his reporters brought him a recording they had bought in the market, and the three listened together. They quickly concluded that the voice on the tape was that of an actor, hired by SAVAK to imitate Khomeini and discredit him. After all, Khomeini was a respected scholar, if a political radical. Why would he stoop to conspiracy theories, telling listeners that the shah had commissioned a painting of the Shia leader Imam Ali with blond hair and blue eyes, signifying the shah's hopes that American Christians would dominate Iran? If this wasn't a joke, then it had to be an attempt to frame and discredit the cleric.[2]

A few months later, Iran's minister of information, Daryoush Homayoun, published an editorial in *Ettela'at*, the country's oldest newspaper, titled "Black and Red Imperialism." A wide-ranging smear of the ayatollah, the article accused Khomeini of colluding with the Soviets (the "red" to conservative Islam's "black"), of being a British spy, and of homosexuality. But Homayoun had underestimated the popularity of the exiled scholar. On January 9, 1978, four thousand students took to the streets and demanded retraction of the article. Iran's powerful army quickly quelled the protest, but killed several students and wounded more in the process.[3]

The death of the students opened a cycle of protest and government overreaction that rapidly destabilized the country. Shia custom requires memorial services, called Arbaeen, forty days after a death. Protests accompanied the services for the dead students, and the shah's troops shot more protesters; that provoked more services, more protests, and eventually general strikes. Scholars estimate that as much as 11 percent of Iran's population participated in these protests, a higher percentage of the population than participated in Russia's or France's popular revolutions.[4] By January 1979, it was the shah who had gone into exile, and a triumphant Khomeini returned to Iran, where more than three million Iranians took to the streets to welcome him.[5] Four months later, a referendum with wide popular support declared Iran an Islamic republic.

Khomeini's quick rise surprised the shah's supporters, who had seen Iran moving away from Islam and toward a secular state, where women had the vote and Iran had strong ties to the West. Khomeini's subsequent brutal consolidation of power surprised students who had supported him, taking seriously his promises of freedom and anti-imperialist democracy, only to see hundreds of the ayatollah's political opponents summarily executed.[6] And the exiled Iranian politicians who had flown from Paris to Tehran with Khomeini were certainly surprised when, two years later, many were dead or in exile again.[7]

But perhaps no one was more surprised than Jimmy Carter. On New Year's Eve 1977, days before students took to the streets of Qom, he had toasted the shah, declaring, "Iran, because of the great leadership of the Shah, is an island of stability in a turbulent corner of the world." Carter's analysis was echoed by the CIA, which dismissed the protests of 1978 in August of that year, asserting, "Iran is not in a revolutionary or even 'pre-revolutionary' situation."[8]

How did the intelligence service of the world's most powerful nation misread the Iranian revolution so badly?

In the waning years of the Cold War, the job of America's intelligence analysts began to shift, becoming vastly more complicated. In earlier decades, analysts had known who the nation's main adversaries were and what bits of information they needed to acquire: the number of SS-9 missiles Moscow could deploy, for example, or the number of warheads each missile could carry. They focused on discovering secrets, facts that exist but are hidden by one government from another. But by the time the Soviet

Union completed its collapse in 1991, as Bruce Berkowitz and Allan Goodman observe in *Best Truth: Intelligence in the Information Age*, the intelligence community had a new role thrust upon it: the untangling of mysteries.[9]

The computer security expert Susan Landau identifies the 1979 Islamic revolution in Iran as one of the first signs that the intelligence community needed to shift its focus from secrets to mysteries.[10] On its surface, Iran was a strong, stable ally of the United States in a conflict-torn region. The rapid ouster of the shah and the referendum that turned a monarchy into a theocracy under Khomeini left governments around the world shocked and baffled.

The 1979 revolution took intelligence agencies by surprise because it was born in mosques and homes, not in palaces or barracks. Even if the CIA was watching Iran closely, it was paying more attention to troop strength and weaponry than to cassette tapes sold in the marketplace. Analysts missed a subtle change in Iranian society: the nation was becoming more connected, both internally and to the outside world, through the rise of new communications technologies.

In their book analyzing the events of 1979, *Small Media, Big Revolution*, Annabelle Sreberny and Ali Mohammadi, who both participated in the Iranian revolution, emphasize the role of two types of technology: tools that let people access information from outside Iran, and tools that let people spread and share that information on a local scale. Connections to the outside world (direct-dial long-distance phone lines, cassettes of sermons sent through the mail, broadcasts on the BBC World Service) and tools that amplified those connections (home cassette recorders, photocopying machines) helped build a movement more potent than governments and armies had anticipated.[11]

The ouster of autocrats in Tunisia, Egypt, and Libya in the 2011 Arab Spring protests has reopened a conversation about the role of technology in enabling social change. Did cassette recorders overthrow the shah? No more than Facebook ousted Mubarak. But in both cases, the technological, political, and social fabric shifted, and old ways of anticipating what changes might occur were no longer applied. Looking for secrets—the missing information in systems we understand—we can easily glide past mysteries, events that make sense only when we understand how systems have changed.

As we enter an age of ever-increasing global connection, we are experiencing vast but subtle shifts in how people communicate, organize themselves, and make decisions. We have new opportunities to participate in conversations that are local and global, to argue with, persuade, and be persuaded by people far from our borders. And we have much to argue about, as our economies are increasingly intertwined, and our actions as individuals and nations affect one another's climate, health, and wealth. And as these connections increase, it should be no surprise that we will also experience a concomitant rise in mystery.

The mysteries brought to the fore in a connected age extend well beyond the realm of political power. Bad subprime loans in the United States trigger the collapse of an investment bank, which tightens interbank lending, pushing Iceland's heavily leveraged economy into collapse, leaving British consumers infuriated at the disappearance of their high-yield savings accounts at Icelandic banks. A family wedding in Hong Kong leaves the World Health Organization tracing a deadly epidemic from Toronto to Manila, the disease spreading as fast as individuals can travel. Not all mysteries are tragedies. Political revolutions, broadcast live from Tunisia, send students into the street in Gabon to demand lower tuition, and inspire labor activists in Wisconsin to seize the state capital. A Korean pop song mocking the materialism of a neighborhood in Seoul, PSY's "Gangnam Style," becomes a global dance hit in an instance of unexpected and convoluted connection.

Uncovering secrets might require counting missile silos in satellite images or debriefing double agents. In order to unwind a banking collapse or combat SARS, we need different skills. Landau suggests that "solving mysteries requires deep, often unconventional thinking, and a full picture of the world around the mystery."[12]

The popular embrace of the Internet means we have a wealth of new ways to learn what's going on in other parts of the world. It's as easy to access the front page of a newspaper from another continent as it is to read one from the next town. In fact, sometimes it's easier. A free online encyclopedia offers background and context on events that would have been hard to obtain ten years ago without visiting a good library. Google promises to organize the world's information and make it universally accessible, and we've grown used to asking it and other search engines to discover secrets: just type "How many SS-9 missiles does the USSR have" and hit "I'm feeling lucky."

These tools help us discover what we want to know, but they're not very powerful in helping us discover what we might need to know. What we want to know is shaped by what, and who, we think is important. We follow the news in our hometowns more closely than news an ocean away, and the lives of our friends in more detail than those of distant strangers. Our media tools, ranging from our newspapers to our social networks, embody those biases; they help us find what we want, but not always what we need.

What do we need to understand a complex and interconnected world? That's not just a question for intelligence agents. Epidemiologists and CEOs, environmentalists and bankers, political leaders and activists are all trying to tackle challenges of global scale. We all need ways to access perspectives from other parts of the world, to listen to opinions that diverge from our preconceptions, and to pay attention to the unexpected and unfamiliar.

We move from unearthing secrets to unwinding mysteries not just through the force of will. Our understanding of the world comes to us through the tools we use to learn about the world around us. Some of those tools are hundreds of years old, whereas others were invented in the past decade. And all of them can be changed to help us better understand and explore the world.

We can build new tools that help us understand whose voices we're hearing and whom we are ignoring. We can make it easier to understand conversations in other languages, and to collaborate with people in other nations. We can take steps toward engineering serendipity, collecting insights that are unexpected and helpful. With a fraction of the brainpower that's gone into building the Internet as we know it, we can build a network that helps us discover, understand, and embrace a wider world.

We can, and we must, rewire.

DISCONNECT

Chapter 1

CONNECTION, INFECTION, INSPIRATION

DR. LIU JIANLUN WASN'T FEELING WELL WHEN HE CHECKED into room 911 of Hong Kong's Metropole Hotel. On February 21, 2003, the sixty-four-year-old medical professor had arrived in Hong Kong to attend a family wedding, but he was feeling exhausted, not festive. For the prior three weeks, he'd been working long shifts at Sun Yat-sen Memorial Hospital in Guangzhou, where an "atypical pneumonia" had sickened hundreds of patients.[1]

Dr. Liu went sightseeing with his brother-in-law, but returned to the hotel early. The next morning, he walked down Waterloo Road to Hong Kong's Kwong Wah Hospital and checked himself in. Gasping for breath, he warned doctors and nurses that he was carrying a highly infectious disease and needed to be treated in a pressurized room.[2]

Ten days later, Dr. Liu died from severe acute respiratory syndrome (SARS). His brother-in-law died soon afterward. Dr. Liu was not the first person to die of SARS, but his case was the first to reveal the potential for the disease to quickly spread over long distances; it eventually claimed a total of 916 lives worldwide during a global epidemic that had the potential to infect billions.[3]

By the time he was isolated in the hospital, Dr. Liu had already

infected twelve other guests staying on the ninth floor of the
Metropole. The infected guests hailed from Singapore, Aus-
tralia, the Philippines, and Canada, as well as from China and
Hong Kong.[4] One of the unlucky ninth-floor guests was Johnny
Chan, an American businessman based in Shanghai. He left the
Metropole two days after Dr. Liu checked in, and flew to Hanoi.
A few days later, he fell ill and was hospitalized at the Vietnam
France Hospital.

When Vietnamese doctors weren't able to diagnose his illness,
they turned to Carlo Urbani, director of infectious diseases for
the Western Pacific Region of the World Health Organization
(WHO). An expert diagnostician, Dr. Urbani quickly determined that whatever was killing Chan was highly contagious. He
immediately met with Vietnamese authorities to ensure that the
country's hospitals took strenuous precautions. But by the time
Dr. Urbani was called in, Chan had already infected eighty other
patients and health care workers at the hospital.

On March 11, the Vietnamese government quarantined the
hospital. Urbani was on a flight from Hanoi to Bangkok to attend
a medical conference. During the flight, he developed a high fever,
one of the few early symptoms of the disease. After getting off
the plane, Urbani isolated himself and called a colleague from the
US Centers for Disease Control and Prevention, Dr. Scott Dowell,
who met him at the Bangkok airport. They talked for almost two
hours, sitting more than eight feet apart, before Thai authorities
could equip an ambulance and medical technicians with sufficient
protective gear to transport Urbani to a hospital.[5] While Urbani
was in good health, doctors believe, he was exposed to the virus
dozens of times from the patients he was treating, overwhelming
his system with a massive viral load. Urbani died on March 29.

If you were in the United States or in Europe when SARS
was discovered, you may have vague memories of travel restrictions and the sudden appearance of hospital masks on foreign
travelers. But only twenty-seven US residents were infected with

SARS, while over seven thousand Chinese contracted the disease. In other parts of the world, the psychological impact of SARS was pronounced. The global health scholar Laurie Garrett notes, "Though most American soon forgot SARS, for many Asians and Canadians the period from November 2002 to June 2003 remains as starkly memorable as the date 9/11 for residents of Washington and New York."[6]

Their concerns were not misplaced: SARS is terrifying. No physical contact is required for disease transmission; you simply need to share airspace with an infected person for an extended period of time. People may incubate the disease for up to ten days without showing recognizable symptoms, which means they can spread the infection over vast distances as they travel, and a single person may infect dozens or hundreds of others. And for roughly one of ten persons infected, SARS was fatal.

During the 2002–03 outbreak, the disease spread with such speed that conspiracy theories formed in its wake. An idea put forward by an obscure Russian scientist became popular in some Chinese chat rooms: any disease this virulent and fast-spreading must be man-made.[7] The truth is weirder, and possibly more disturbing. By April 2003, WHO scientists had discovered that SARS was a virus native to the masked palm civet, a catlike carnivore common in southern China. Like ebola, anthrax, and hantavirus, SARS is zoonotic; it spreads from animals, who may carry a disease without suffering from it, to humans. SARS likely crossed the species barrier through the blood of civets, sold for meat in the markets of southern China, and then passed human to human from the civet eaters to Drs. Liu and Urbani.

With its long incubation period and ease of transmission, SARS seemed custom-made for a connected world. "Super spreaders" like Dr. Liu and Johnny Chan, mobile professionals who traveled by airplane between major global cities, took the disease with them. On a single flight—China Airways 112 from Hong Kong to Beijing on March 15—a single passenger infected

22 of 126 fellow passengers.[8] As fear spread, people grew anxious about the dangers of airplanes, public transportation, and the other shared spaces of global megacities. Anonymous copresence with thousands of others—a routine experience of modern urban life—suddenly appeared unreasonably risky. Like Edward Lorenz's butterfly, whose wing beats in Brazil set off tornadoes in Kansas, someone's dinner in China suddenly had the potential to cause hospitalizations in Canada.

SARS eventually reached thirty-two countries and every continent save Antarctica, but only 8,422 people fell ill. And although it spread explosively from November 2002 to March 2003, by July 2003 the WHO could confidently declare the epidemic contained.[9] In the end, the most interesting thing about the SARS outbreak was not how fast the disease spread but how quickly it was stopped.

Contrast the SARS numbers with an earlier epidemic, the Spanish flu. From 1918 to 1920, one-third of the world's population, roughly half a billion people, contracted a deadly form of influenza, and an estimated 50 million died.[10] Case for case, the Spanish flu was less deadly than SARS. About 2.5 percent of Spanish flu cases were fatal, though many persons caught it more than once. SARS killed 9.6 percent of those it infected and was especially deadly for the elderly, who had a mortality rate of more than 50 percent. The Spanish flu, like SARS, was mobile; outbreaks occurred on remote Pacific islands and above the Arctic Circle. But the Spanish flu sufferers who carried the disease to these far corners traveled by steamship and train, not transoceanic jet. Why did SARS, a disease so deadly, so well positioned to spread globally, kill so few people?

One big part of the answer is the Internet. Global cooperation and communication stopped SARS, and the ability of doctors around the world to connect and collaborate online made the Internet the front line for stopping the disease.

When Dr. Urbani, the Italian diagnostician, put the Vietnamese government on alert in March 2003, he triggered a global effort by the WHO to identify, diagnose, and contain SARS. Six days after Urbani landed in Bangkok, the WHO rolled out a secure website that hosted videoconferences to coordinate the efforts of researchers in labs around the world. They shared lung x-rays of infected patients to develop a diagnostic protocol, which they then disseminated to hospitals around the world, along with guidelines for isolating infected patients. The alerts proved remarkably effective—90 percent of all the SARS cases occurred before the WHO's advisories were issued. To identify ongoing or new SARS outbreaks, the WHO used the Global Public Health Intelligence Network (GPHIN), a software tool developed by Canada's national health ministry that scans newswires and Internet sources for mentions of possible SARS outbreaks or other unexplained health events. More than one-third of the rumors identified by GPHIN led the WHO to identify and isolate cases of SARS.[11]

The WHO monitored newspapers and social media in part because not every national government issued accurate reports about the spread of the disease. China was profoundly affected by SARS and, not coincidently, was also the country that was most reluctant to share information about infections. More than two weeks after the WHO's global alert, Chinese officials were publicly claiming that Beijing had experienced only twelve cases of SARS. Dr. Jiang Yanyong, a Beijing-based doctor, had personally treated fifty SARS patients and knew those numbers were artificially low. He emailed TV stations in Beijing and Hong Kong with his concerns, and his email was passed to reporters at the *Wall Street Journal* and *Time* magazine, who brought international attention to his claims.[12] Less than two weeks after *Time* published a cover story on SARS in Beijing, the Chinese national health minister and the mayor of Beijing were fired. The new mayor closed schools, discos, and theaters, following instructions from the WHO. International attention and scrutiny had brought China quickly onto the global team combatting SARS.

The ability to share information without sharing the same air helped minimize the disruptions that SARS and resulting quarantines caused. Singapore, one of the nations first affected by SARS, isolated SARS patients in a single ward, then released them after treatment into home quarantine, monitored by government-installed videoconferencing units. In a truly inventive turn, the Singapore government also discouraged the local Chinese population from celebrating Ching Ming, a holiday in which believers assemble in cemeteries and clean their ancestors' graves. Anxious not to create crowds in the city's cemeteries, the government encouraged residents to buy offerings through an online service that arranged for a uniformed attendant to clean a grave and make the offering on their behalf.[13]

Writing about the WHO's success in containing SARS, Dr. Shigeru Omi, regional director for the western Pacific for WHO, speculates that SARS would never have expanded beyond a small, regional outbreak if not for international jet travel, and that the WHO wouldn't have fought it off so successfully without the Internet as an ally. If international connection through air travel helped spread infection, digital connections—local and international—helped spread the ideas required to fight it. Whether through doctors on different continents jointly examining x-rays, or officials in Toronto and Singapore discussing quarantine strategies, connection can inspire crucial collaborations just as well as it can spread infection.

Epidemics unfold like mysteries. We don't know where in the world they will emerge, or what previously harmless and commonplace practices will spread them across the globe in a single day. To diagnose and stop epidemics, scientists like Dr. Omi need to follow leads locally and globally. A broad view of the world is essential if they are to identify potential threats and embrace creative solutions. The GPHIN network that helped WHO researchers collect tips and rumors from newspapers and online media was powerful precisely because it was looking for SARS not just in China and Hong Kong but in all corners of the world.

SARS offers one example of the global-scale challenges we face today. There are many more, including a rapidly changing climate, interconnected and teetering financial systems, and competition for arable land and other scarce natural resources. Optimism permits us to imagine a wave of networks like GPHIN scanning the horizons for threats and opportunities and speeding out responses but one still unfolding mystery suggests that the view of the horizons that we need remains obscured.

. . . . . .

HAD YOU ASKED A GROUP of experts on the Middle East in 2010 what changes they thought were likely to take place in the following year, almost none of them would have predicted the Arab Spring movement. Certainly not a single one would have chosen Tunisia as the flash point for the events that followed. Zine el-Abidine Ben Ali had ruled the North African nation virtually unchallenged since 1987 and had co-opted, jailed, or exiled anyone likely to challenge his authority. When the vegetable seller Mohamed Bouazizi set himself on fire in December 2010, there was no reason to expect his family's protests against government corruption to spread beyond the town of Sidi Bouzid.[14] After all, the combination of military cordons, violence against protesters, a sycophantic domestic press, and restrictions on the international press had, in the past, ensured that dissent remained local.

Not this time. Video of protests in Sidi Bouzid, shot on mobile phones and uploaded to Facebook, reached Tunisian dissidents in Europe. They indexed and translated the footage and packaged it for distribution on sympathetic news networks such as Al Jazeera. Widely watched in Tunisia, Al Jazeera alerted citizens to protests taking place in other corners of their country. The broadcasts also acted as an invitation to participate. Ben Ali took to the airwaves, alternately begging protesters to disperse and threatening them if they didn't. As his regime trembled and fell, images of the protests spread throughout the region, inspiring similar outpourings in more than a dozen countries and the eventual overthrow of Hosni Mubarak in Egypt and Muammar Gaddafi in Libya.

Although the impact of Tunisia's revolution is now acknowledged, at the time they occurred the protests that led to Ben Ali's ouster were invisible to much of the world. The *New York Times* first mentioned Mohamed Bouazizi and Sidi Bouzid in print on January 15, the day after Ben Ali fled the country.[15] The Lebanese American journalist Octavia Nasr had followed the story from early on and expressed her frustration in an interview with PBS: "For four weeks, Tunisia was ignored in our media. They didn't pay attention to the story until it was so huge and in their face, they couldn't ignore it anymore."[16]

Some observers suggested that the silence of American and European media reflected government support for Ben Ali: so long as the United States considered Ben Ali a useful ally, media outlets weren't inclined to report on the story. While thoughtfully cynical, this scenario fails to explain why the movement to overthrow Mubarak, a close ally central to US interests in the region, received widespread coverage in American media, whereas the Tunisian revolution registered only when it was over.

Here is a simpler, conspiracy-free explanation: most Americans and Europeans missed the Tunisian revolution because they weren't paying attention. The protests gained real momentum over Christmas and New Year's, a time when many people focus their attention on family and friends instead of news of the world. Tunisia's government-dominated press didn't report on the protests, and independent media sites tracking the events were largely unknown outside the Tunisian diaspora.

As it turns out, the US intelligence community wasn't paying much attention either. President Obama later confronted

National Intelligence Director James Clapper and told him he was "disappointed with the intelligence community" for its failure to provide adequate warning of the ouster of the Ben Ali and Mubarak governments. Senator Dianne Feinstein, who chairs the US Senate Intelligence Community, wondered why protests that had spread in large part because of social media escaped the scrutiny of military intelligence: "Was someone looking at what was going on [on] the Internet?"[17]

Whether we're concerned with fighting epidemics like SARS or reacting to geopolitical shifts like the Arab Spring, we need a broad, global picture so that we can anticipate threats, seize opportunities, and make connections. The existence of mobile telephony, satellite television, and the Internet suggests that information should be available from throughout the world at unprecedented volumes. Yet a central paradox of this connected age is that while it's easier than ever to share information and perspectives from different parts of the world, we may now often encounter a narrower picture of the world than in less connected days.

Four decades ago, during the Vietnam War, reporting from the front lines involved transporting exposed film from Southeast Asia by air, then developing and editing it in the United States before broadcasting it days later. In our era, an unfolding crisis, be it a natural disaster or a sudden military coup, can be reported in real time via satellite. Yet, despite these lowered barriers, US television news today features less than half as many international stories as were broadcast in the 1970s.[18]

With more than two billion people connected to the Internet and six billion people with access to mobile phones,[19] weather reports from rural Mali or reports on the local politics in Bihar are more easily retrieved today than at any time in the past. Our challenge is not access to information; it is the challenge of paying attention. That challenge is made all the more difficult by our deeply ingrained tendency to pay disproportionate attention to phenomena that unfold nearby and directly affect ourselves, our friends, and our families.

In *Six Degrees*, his exploration of networked phenomena like epidemics, fads, and financial crises, the mathematician Duncan Watts argues that our lives are affected by phenomena that are geographically distant from us. "Just because something seems far away, and just because it happens in a language you don't understand, doesn't make it irrelevant," he argues. "To misunderstand this is to misunderstand the first great lesson of the connected age: we may all have our own burdens, but like it or not, we must bear each other's burdens as well."[20]

This task of bearing one another's burdens forces us to reconsider how we learn about the rest of the world, how we plan strategies and make decisions, how we build our businesses, govern our nations, and educate our youth. None of these changes are simple, but all start from a simple premise. We must begin to understand ourselves not just as citizens of a state or a nation but also as citizens of the world. This is not a new idea, of course. One of its earliest recorded expressions can be found in the life of a Greek man born in the fourth century BC.

Cosmopolitanism

For a guy who could travel only on foot or by ship, Diogenes managed to see a large fraction of the world known to his people. Exiled from his native Sinope (on the Black Sea coast of contemporary Turkey),[21] Diogenes found himself living penniless on the streets of Athens, and later in Corinth. Following the teachings of Socrates's student Antisthenes, Diogenes became an ascetic, which was probably an excellent career move, since he had already been relieved of his earthly wealth. Accounts of his life are sparse and resemble legend as much as history, but most classicists agree that Diogenes was homeless and slept in a wooden tub under the awnings of Athenian temples.

In his *Lives and Opinions of the Eminent Philosophers*,[22] Diogenes comes off as a cross between Woody Allen and Old Dirty Bastard, delivering memorable quips and behaving inappropriately. Found masturbating in the agora, Diogenes didn't apologize for his behavior, but noted that he wished it were similarly easy to relieve his hunger by rubbing his belly. Termed "the dog" by some contemporaries (the Greek word for dog, κύων, is the root of the term "cynic"), Diogenes reacted to being tossed food scraps at a banquet as any dog would: by urinating on his benefactors. While many historians see him as a philosophical innovator and an important critic of Plato, others view him as a colorful madman.

For all his outrageous behavior, Diogenes is best known for his refusal to identify as an Athenian or a Sinopean. Instead, he declared that he was a citizen of the universe, a cosmopolitan: (κόσμος, universe, πόλις, city). Diogenes's declaration of cosmopolitanism was hardly representative of mainstream classical Greek thought. Instead, it was a radical assertion. Virtually everyone in Diogenes's universe identified closely with the city-state in which he was born and resided. Diogenes was probably not embracing a global identity as much as he was rejecting the key social identifier of his age: a man's place of origin.

Challenging as Diogenes's statement was to his contemporaries, it has always been easier to declare ourselves cosmopolitans than to actually live in a wide world.

We are 2,500 years removed from Diogenes, but it's still only in very recent times that the majority of people have had the opportunity to interact with people from different parts of the world. In 1800, some 97 percent of the world's population lived in rural areas.[23] While some people may have had contact with cultures through visiting merchants or other travelers, most never met anyone who spoke a different language or worshiped a different god. The 3 percent who lived in cities like Athens before 1800 had the rare opportunity to talk to, trade with, and worship alongside people who had different origins, languages, and gods. Although those early cities were the first spaces where lived cosmopolitanism was actually possible, we probably overestimate the degree of cultural mixing that occurred in them.

The historian Margaret Jacob recently studied eighteenth-century descriptions of the stock exchanges in Europe's most cosmopolitan cities. Jacob observes that though in contemporary accounts traders from throughout Europe and beyond participated, divisions between groups were definite: "A 1780s manuscript sketch of the floor of the London exchange made by a visiting French engineer suggests that by then national identities competed with professional as well as religious ones. The sketch of the floor plan shows the familiar groupings—'place hollandaise,' 'place des Indes Orientales,' 'place Française'—but also new ones: 'the place of the Quakers,' 'the place of the Jews.' . . ." The traders may have been Londoners working in the greatest global marketplace of the time, but their origins and faiths formed their primary identity.

As described, the eighteenth-century London stock market sounds curiously like today's multicultural cities. Consider New York, where residents know that Brighton Beach is home to thousands of Russian speakers, Flushing to a large Chinese community, Borough Park to Orthodox and Hassidic Jews. The promise of our contemporary cities is that it's possible to encounter different foods, customs, and ideas through incidental encounters with neighbors or through the conscious decision to take a subway ride to a different corner of town. But how often does this happen? "Lived cosmopolitanism," Jacob notes, has always been more difficult than merely building spaces where people from different backgrounds come together.[24]

In 2006, the celebrated social theorist Robert Putnam shared results from his Social Capital Benchmark Survey, which suggest that contemporary Americans have a long way to go before they embrace the multicultural promise of a city like New York. In Putnam's study, "people living in ethnically diverse settings appear to 'hunker down.'"[25] They are less likely to vote, work on community projects, give to charity, or volunteer than Americans in less diverse cities. They have less confidence in government's ability to solve problems, fewer friends, and a lower perceived quality of life.

Earlier sociological theories suggested that contact between ethnic groups led either to improved social relations or to conflict between groups—"contact theory" versus "conflict theory." Putnam believes that survey data from American cities point to a third possibility: "constrict theory," a tendency to shy away from contact when presented with diversity.[26] If Putnam's constrict theory is right, and if it also underpins our behavior online, it raises uncomfortable questions about the potentials and realities opened by the Internet. Connecting with people from other backgrounds is hard, even when they live next door or in the same city; paying attention to the problems and concerns of people in the rest of the world is harder still.

The Ghanaian American philosopher Kwame Anthony Appiah has had to think through the possibilities and challenges inherent in cosmopolitanism. Raised between Kumasi and London, the child of a British art historian and a Ghanaian politician, Appiah has explained the intricacies of Ashanti belief systems to Western philosophers and his identity as a gay man to his relatives in Kumasi. Cosmopolitanism, Appiah argues, is about much more than learning to tolerate those who have values and beliefs that differ from ours. We might tolerate practices that offend us by ignoring or turning away from them, but merely tolerating other practices could lead to the hunkering down Putnam warns us about, in which encountering difference makes us constrict our encounters with the world. Instead, Appiah celebrates cosmopolitanism, which, by contrast, challenges us to embrace what is rich, productive, and creative about this difference.

For Appiah, cosmopolitans have two essential qualities. They take an interest in the beliefs and practices of others, striving to understand, if not accept or adopt, other ways of being. In his words, "Because there are so many human possibilities worth exploring, we neither expect nor desire that every person or every society should converge on a single model of life."[27] Cosmopolitans also take seriously the notion that they have obligations to people who are not their kin, even to people who have radically different beliefs. We have obligations to witness and document harms that others suffer from, to lend what assistance we are able, and to treat the people we encounter, no matter how different they are, as part of an extended family.

This two-part definition means that my taste for sushi and my fondness for Afropop are insufficient to make me a cosmopolitan. Appiah saves the label for those who take seriously, and presumably those who act on, their obligations to the people and peoples responsible for the food and the music. Nor is cosmopolitanism a simple universal love of humanity, especially when expressed as a desire to "save" others through Christianity, Islam, democracy, or any other proselytizing faith. We are challenged to take seriously the idea that other possibilities are worth our time and consideration, not our immediate opposition and rejection.[28] When we embrace these, the effect can be uncomfortable and disconcerting. At the same time, it can also be a powerful force for those seeking insight or inspiration.

How We Know the World

In the spring of 1907, Pablo Picasso was visiting Gertrude Stein in her apartment in Paris. Henri Matisse stopped by to show off an African sculpture he had purchased from the Paris dealer Emile

Heymenn, a mask made by the Dan people of western Côte d'Ivoire.[29] Picasso was fascinated by the piece and soon thereafter dragged his friend André Derain to the Trocadero Museum of Ethnology, Paris's first museum dedicated to anthropology. Initially, Picasso was put off: "A smell of mould and neglect caught me by the throat. I was so depressed that I would have chosen to leave immediately."[30] Three decades later, Picasso described himself as haunted by the sight and smell of "that awful museum."[31]

Fortunately for the future of painting, Picasso overcame his initial aversion. He recalled, "I forced myself to stay, to examine these masks, all these objects that people had created with a sacred, magical purpose, to serve as intermediaries between them and the unknown, hostile forces surrounding them, attempting in that way to overcome their fears by giving them colour and form. And then I understood what painting really meant."[32]

His visit to the Trocadero marked the beginning of what Picasso called his "période nègre"—his African period. Later that year, he produced one of his masterpieces, *Les Demoiselles d'Avignon*, a striking portrait of five female nudes, two with faces that closely resemble West African masks. Picasso became a collector of African art, lining the walls of his studio with masks and figurines, and he included African themes in paintings like *Musician with Guitar* (1972),[33] produced shortly before his death. Scholars track Picasso's technique of reversing concave and convex lines in faces, and the transformation of smooth surfaces into geometric solids—the basis of cubism—to his African inspirations.[34] Picasso's appreciation of African art led him into dialogue with leading African intellectuals, including Léopold Senghor, the first president of postcolonial Senegal. Senghor recognized Picasso's showcasing of African themes, and his support for African independence, with a poem dedicated to the painter, "Masque nègre," in his first collection of poems, *Chants d'ombre*.[35]

Picasso found his connection to African influences first through an encounter at a distance, mediated through a museum in

France. Only after wrestling with his reactions to African media did he begin a dialogue with individual Africans like Senghor.

It's tempting to imagine Matisse, projected into the present day, sharing pictures of the Dan mask he'd just bought on Facebook, leading Picasso to frantically search Google for related images. We are less likely to find our connections to the unfamiliar—the infectious and the inspirational—in the physical world. We will likely find them on the screen.

The University of California at San Diego researchers Roger Bohn and James Short calculate that Americans receive information 11.8 hours per day, once we consider information received by broadcast, video, print, telephone, computer, electronic games, and recorded music.[36] Some small fraction of that time is spent encountering news.[37] We spend an increasing percentage of our time on social media, keeping up on the minutiae of friends' and family's lives and movements. Facebook alone now claims 13 minutes per day of a user's time, on average. The remaining hours are spent entertaining ourselves, with music, television, and YouTube videos of cute cats.

Our encounters with these three types of media—news, social media, and cultural media—shape what we know and value. If we keep hearing about a person, place, or event, we register that what we've learned about is important, and we're predisposed to pay attention to the topic. And while one of the great promises of the Internet is that we might encounter anything online, in practice much of what we encounter comes from much closer to home.

We've built information tools that embody our biases toward events that affect those near and dear to us. Our newspapers and broadcasters pay more attention to local and global matters than to international ones. We lean toward television and movies in our native languages, and toward keeping up with high school friends on Facebook instead of using social networks to befriend strangers. Even though Nigerian films and Indonesian news are available through powerful indexes like Google, those tools suffer from another bias: they give us the information we want, not necessarily the information we might need.

What these biases mean, cumulatively, is that we must work hard for our Picasso moments, the moments when an unexpected encounter with the unfamiliar leads to inspiration. And we must work equally hard to build tools that warn us of the dangers of connection, whether it's an incipient epidemic, a financial crisis, or an inflammatory video. The Internet will not magically turn us into digital cosmopolitans; if we want to maximize the benefits and minimize the harms of connection, we have to take responsibility for shaping the tools we use to encounter the world.

It Was Supposed to Be So Easy

In 1993, Howard Rheingold published *The Virtual Community*, reflections on the time he'd spent in early electronic forums, including Internet Relay Chat (IRC), a text-based, real-time chat system created in 1988, but still popular today in technical circles. With chapter titles like "Real-time Tribes" and "Japan and the Net," the book offers the hope that online dialogues will be more fair, more inclusive, and more global than those we've known before. "Thousands of people in Australia, Austria, Canada, Denmark, Finland, France, Germany, Israel, Italy, Japan, Korea, Mexico, the Netherlands, New Zealand, Norway, Spain, Sweden, Switzerland, the United Kingdom, and the United States are joined together at this moment in a cross-cultural grab bag of written conversations known as Internet Relay Chat (IRC)." Rheingold wonders, "What kinds of cultures emerge when you remove from human discourse all cultural artifacts except written words?"[38]

Rheingold was not the first to hope that an emerging technology would transform the ways distant strangers relate to one another. In his book *The Victorian Internet*, Tom Standage, the

Economist's technology editor, offers a compendium of optimistic predictions for the telegraph, or "the highway of thought," as one contemporary commentator called it. In one of Standage's many examples, the completion of a submarine cable linking the United States and the United Kingdom moved the historians Charles Briggs and Augustus Maverick to assert, "It is impossible that old prejudices and hostilities should longer exist, while such an instrument has been created for the exchange of thought between all the nations of the earth."[39]

The arrival of the airplane inspired similar rhetoric. Commenting on Louis Blériot's crossing of the English Channel in 1909, the *Independent* of London suggested that air travel would lead to peace because the airplane "creates propinquity, and propinquity begets love rather than hate." A similar logic led Philander Knox, US secretary of state under the president Howard Taft, to predict that airplanes would "bring the nations much closer together and in that way eliminate war."[40]

Interviewed in 1912, the radio pioneer Guglielmo Marconi declared, "The coming of the wireless era will make war impossible, because it will make war ridiculous."[41] Even after the Great War had rendered Marconi's pronouncement absurd, the inventor Nikola Tesla saw an even grander future for radio: "When wireless is perfectly applied the whole earth will be converted into a huge brain. . . . We shall be able to communicate with one another instantly, irrespective of distance."

As befits a man of his genius, some elements of Tesla's 1926 vision were surprisingly accurate. "Through television and telephony," he said "we shall see and hear one another as perfectly as though we were face to face, despite intervening distances of thousands of miles; and the instruments through which we shall be able to do his will be amazingly simple compared with our present telephone. A man will be able to carry one in his vest pocket."[42]

These and other observations will sound familiar to anyone

who witnessed the rise of the Internet. As the historian and technology scholar Langdon Winner suggests, "The arrival of any new technology that has significant power and practical potential always brings with it a wave of visionary enthusiasm that anticipates the rise of a utopian social order."[43] Technologies that connect individuals to one another—like the airplane, the telegraph, and the radio—appear particularly powerful at helping us imagine a smaller, more connected world. Seen through this lens, the Internet's underlying architecture—it is no more and no less than a network that connects networks—and the sheer amount written about it in the past decade guaranteed that the network would be placed at the center of visions for a world made better through connection. These visions are so abundant that they've even spawned a neologism: "cyberutopianism."

The term "cyberutopian" tends to be used only in the context of critique. Calling someone a cyberutopian implies that he or she has an unrealistic and naïvely overinflated sense of what technology makes possible and an insufficient understanding of the forces that govern societies. Curiously, the commonly used term for an opposite stance, a belief that Internet technologies are weakening society, coarsening discourse, and hastening conflict is described with a less weighted term: "cyberskepticism." Whether or not either of these terms adequately serves us in this debate, we should consider cyberutopianism's appeal, and its merits.

In a Skype conversation with Howard Rheingold, I mentioned that I planned to include some of his thoughts in this book's discussion of cyberutopianism. On being linked to the term, Rheingold was flustered, and I briefly thought he might hang up on me. Instead, he paused, composed himself, and offered the observation that "the Abolitionists were utopians." In a later email he explained further,

I am enthusiastic about the potential for tools that can enhance collective action, but as I stated on the first page of Smart Mobs [his 2002 book on technology and collective action], humans do beneficial things together and they do destructive things together, and both kinds of collective action are amplified. . . . So although I recognize that Communism and Fascism were sold as utopias, I like to reverse my logic—not only do people do really bad things under utopian banners, they can also do things like move for the abolition of slavery.[44]

Rheingold's comment reminds us not to let our opponents frame the debate. "Cyberutopianism" is an uncomfortable label because it combines two ideas worthy of careful consideration into a single, indefensible package. The belief that connecting people through the Internet leads inexorably to global understanding and world peace is one not worth defending. Believing that technologies influence whom and what we know and care about is a more complicated idea, and one worth our close consideration. As with Appiah's concept of cosmopolitanism, it's not enough to be enthusiastic about the possibility of connection across cultures, by digital or other means. Digital cosmopolitanism, as distinguished from cyberutopianism, requires us to take responsibility for making these potential connections real.

If we reject the notion that technology makes certain changes inevitable, but accept that the aspirations of the "cyberutopians" are worthy ones, we are left with a challenge: How do we rewire the tools we've built to maximize our impact on an interconnected world? Accepting the shortcomings of the systems we've built as inevitable and unchangeable is lazy. As Benjamin Disraeli observed in *Vivian Grey*, "Man is not the creature of circumstances, circumstances are the creatures of men. We are free agents, and man is more powerful than matter."[45] And, as Rheingold suggests, believing that people can use technology to build a world that's more just, fair, and inclusive isn't merely defensible. It's practically a moral imperative.

Building for a Wider Future

Cyberutopianism offers us the reassurance that technological innovations will lead to social progress, to positive connections between people with different perceptions and beliefs. But the case of SARS suggests that connection is a double-edged sword, opening us to both the danger of infection and the potential of new solutions. A recent YouTube video helps demonstrate just how hard we might need to work in order to turn encounter across culture into the positive force that digital cosmopolitanism suggests it could be.

In the summer of 2011, the filmmaker Sam Bacile recruited actors for a film titled *Desert Warriors*. Wearing turbans, flowing robes, and sandals, the actors performed in front of a green screen, in an industrial space in Monrovia, California, far from the studios of Hollywood. The convoluted plot involved battles between warring tribes, provoked by the arrival of a comet. The script was so poorly written that the actors made fun of it between takes, and the director didn't seem to care if the actors flubbed their lines, moving quickly from scene to scene.[46]

On July 1, 2012, Bacile posted a fourteen-minute trailer for his movie, now titled *Innocence of Muslims*, on YouTube. Watching the trailer makes it clear why the director wasn't concerned with the actors' delivery: the film was obviously dubbed, and the actors now delivered lines about the Prophet Muhammad, portraying the Prophet as a sex-obsessed, violent pedophile. The film attracted little notice among YouTube audiences, garnering only a few thousand views, but caught the attention of two vocal opponents of Islam, Pastor Terry Jones and the Coptic activist Morris Sadek.

Jones and Sadek both have long records of anti-Islamic provocation. Jones is best known for launching "International Burn a Quran Day" on the ninth anniversary of the September 11 attacks. His plans to burn Islam's holy book led to protests in the United States and abroad, widespread media coverage and meetings between Jones and senior US officials, who persuaded him not to carry out his threats.[47] Jones promoted Bacile's film to his followers as part of his September 11, 2012, "International Judge Muhammed Day," a sequel to his earlier Quran-burning plans. Sadek, known to the Coptic community for his frequent emails denigrating Islam, posted Bacile's video, with Arabic subtitles, to the website of his group, the National American Coptic Assembly. He also sent hundreds of emails promoting the video to colleagues in Egypt.[48]

The *Innocence of Muslims* film eventually came to the attention of the Egyptian TV host Sheikh Khaled Abdullah. Abdullah appears on Al-Nas television, a satellite channel based in Cairo, known for its conservative Islamic stances. For religious reasons, Al-Nas has no female presenters, and when Abdullah aired the clip on September 8, condemning it as an American attack on Islam, the faces of women in the video were blurred out. The video had been dubbed into Arabic, which made it impossible to tell that the English-language audio had been cut and pasted together, and Abdullah and other commentators implied that the film had been sponsored or supported by the US government and shown on "state television" in the United States.[49]

Al-Nas is watched throughout the Arabic-speaking world, and audiences in Egypt and Libya responded to the broadcast by protesting at American embassies in Cairo and Benghazi on September 11, 2012. In Cairo, protesters breached the outer wall of the embassy, tore down American flags, and hoisted black flags with text declaring, "There Is No God But God and Allah Is His Prophet."

The damage in Benghazi was much more serious. Angered by the video, the Islamist militia Ansar al-Shariah attacked the US consulate and set it on fire, trapping Ambassador Christopher Stevens and others inside. Stevens and four others died of smoke inhalation. Despite strong condemnations of the video from President Obama and security crackdowns across the Middle East, protests against the film erupted in Somalia, Pakistan, Sudan, and locations as far flung as Australia and Belgium.[50]

Violent protests were, of course, what the filmmaker had wanted. "Sam Bacile" is believed to be Nakoula Basseley Nakoula, an Egyptian American Coptic Christian with a criminal past. Nakoula's initial target audience for his film was Muslims living in Los Angeles. He screened the film in a rented theater in Hollywood on June 23, 2012, as *The Innocence of Bin Laden* and tried to attract an audience by running an Arabic-language advertisement in local newspaper, hoping those inclined to believe in Bin Laden's innocence would attend.[51] Nakoula failed to provoke a reaction from local Muslims, because few, if any, attended the screening. But Jones and Sadek worked to make sure that a wider audience would see the film and take offense. And given that Jones and Sadek argue that Islam is a dangerous religion, the burning of the Benghazi consulate represents a victory, proof positive for their equation that Islam equals violence.

In web parlance, Nakoula, Sadek, and Jones are trolls, persons who try to hijack a discussion with harassing or inflammatory content, hoping to provoke a response. The Internet scholar Judith Donath notes that "the troll attempts to pass as a legitimate participant, sharing the group's common interests and concerns," but his ultimate intention is not to engage in discourse but to incite.[52] Because trolls need to disguise themselves, it requires some skill to troll a discussion successfully. Immediately posting insults ends a conversation quickly, while morphing from a legitimate commentator to a provocateur yields the anger the troll desires. Over time, Internet users have developed some resistance to trolls. "Don't feed the trolls" is a standard admonition—in other words, if someone is trying to incite you, don't bother responding. The broader media ecosystem, however, has yet to develop robust troll defenses.

It may be obvious to a Western viewer that the sole purpose of *Innocence of Muslims* is to provoke, but it's less obvious when dubbed into Arabic and presented as a new film made for and aimed at mainstream American audiences. We can think of the movie as an infection designed to exploit the predispositions of our media systems.

At the same time that some media players in the Middle East are actively searching for evidence that the United States is persecuting Muslims, the US media since 9/11 have paid disproportionate attention to violence committed by Muslims. The protests played into an existing narrative for American news outlets, a narrative best illustrated by the cover of *Newsweek*'s September 24, 2012, issue, which features an image of bearded men in turbans yelling, under the headline "Muslim Rage."

The trolls behind *Innocence of Muslims* exploit both of these narratives. They provide Middle Eastern Muslims evidence that Americans misunderstand and disrespect Islam so badly that hundreds of people were willing to get together and make a film that insults the Prophet. The ensuing protests play to American media's focus on the sudden and violent, at the expense of processes that may be more important, but are hard to portray visually: the authoring of a Libyan constitution, peaceful elections in Egypt. *Newsweek*'s cover invites us to see the Libyan protest the way Nakoula and Pastor Jones see it, as evidence that Islam is unpredictable and violent. Appiah's vision of cosmopolitanism suggests we look more deeply and see whether the situation is more complex than it appears at first glance. With a little work and a wider view, a very different narrative emerges.

Marc Lynch, a leading scholar of Arab media, points out that the protests, while sometimes violent, "were actually quite small—vastly inferior in size and popular inclusion to the Arab uprisings protests last year, and small even in comparison to the ongoing pro-democracy or other political demonstrations which occur on a weekly basis in many Arab countries." One protest that wasn't widely covered took place on September 21, ten days after the US consulate in Libya burned, "where tens of thousands came out in Benghazi in an inspiring rally against militias and against the attack on the U.S. consulate."[53] A day later, similar rallies ousted the Ansar al-Sharia militia, which is believed to have set the US consulate on fire, from their base near the city.[54]

While dozens of op-ed writers picked up their pens to opine on Muslim rage, Lynch notes, few have been inspired to write about these massive rallies in support of the United States. The op-ed writers Lynch is waiting for may find inspiration in a YouTube video that offers a very different window on the Libyan protests. Shot by the Libyan activist Fahd al-Bakoush, it shows a dozen men carrying Ambassador Stevens, unconscious from smoke inhalation, out of the burning consulate to a car, to take him to the hospital. When the men discover Stevens is still alive, they chant, "God is Great."[55]

While tens of thousands of Benghazi residents marched against one manifestation of "Muslim rage," American Muslims reacted to the *Newsweek* cover in a more subtle and snarky fashion. To accompany its story, *Newsweek* invited people to share their thoughts online, using the Twitter hashtag #muslimrage. Hundreds of Muslims in the United States and elsewhere did so, posting pictures of themselves looking mildly annoyed, with captions that show their "#muslimrage" at the frustrations of ordinary life. A collection of these images, at http://muslimrage.tumblr.com/, features captions like these:

My bookmark fell out and now I'ma have to page through to find my spot. #MuslimRage

kebabs burning! why my timer didn't go off? #MuslimRage

3-hour lecture tomorrow at 8 am. Why. #MuslimRage

The #muslimrage tweets point up an obvious truth: the violent protesters represent an infinitesimal fraction of the nearly two billion Muslims worldwide.[56] Most Muslims don't look like the scary men on the cover of *Newsweek*; rather, they look like friends, neighbors, colleagues, and classmates, and their frustrations, for the most part, are the petty, everyday ones we all share.

With marches in Benghazi and tweets in the United States, Muslims are trying to fight a simplistic narrative that obscures the larger transformation taking place in the Middle East—a move from a world of oppressive autocrats and suppressed religious movements to representative governments that try to balance moderate Islam and electoral democracy. Our inability to see the smiling and sarcastic #muslimrage because we're blinded by the overblown and violent "Muslim rage" suggests that we're getting a distorted picture of the world. This limited view, attuned to some narratives and not to others, makes it hard to anticipate and understand major shifts like the Arab Spring.

We cannot escape a connected world. Governments will block access to YouTube in the wake of the *Innocence of Muslims* trailer, just as they grounded planes during the SARS outbreak. But ideas, both the ugly and the inspiring, will spread across borders.

To succeed in a connected world, to fight infection and embrace inspiration, we need a wider view. We need to encounter unexpected influences, like the masks that shaped Picasso's career. We need to put events like the Benghazi attack in proportion, and we need to discover what's missing. We need to take a longer and wider look, approaching the first explanations of a mystery with skepticism, probing for a fuller picture. We need to find guides who can help us translate and contextualize what we're seeing so that we can understand what's really going on in the world.

A future of connection across lines of language, culture, and nation is made more possible by the rise of the Internet. Our economic and creative success depends on our becoming digital cosmopolitans, on embracing inspirations and opportunities from all parts of the world. To build the tools we need to thrive in this emerging world, we must understand how we're connected and disconnected.

We need to move toward a physics of connection, an understanding of what's necessary to build real and lasting connections in digital space. Our first step toward that goal is establishing a better understanding of what we actually do, and don't do, and whom we hear, and don't hear, when we use the Internet. We have to take a close look at how connected we are, not just at how connected we imagine ourselves to be.

Chapter 2

IMAGINARY COSMOPOLITANISM

MIT'S PROFESSOR NICHOLAS NEGROPONTE HELPED BRING the Internet into the public spotlight with his 1995 book, *Being Digital*, which celebrated a near-future world in which digital technologies transformed every aspect of our lives. For a book that touches on holographic video, virtual reality, and other as yet unrealized aspects of life on the net, it starts in a surprisingly mundane place. Attending a conference on American competitiveness, Negroponte noted the irony of being served Evian water, shipped in glass bottles from the French Alps. America's future, he declared, was not in manipulating these heavy, hard-to-move atoms, but in weightless bits.[1]

There's no Evian water for sale at the convenience store around the corner from MIT's Media Lab, the interdisciplinary technology research center that Negroponte founded in 1985 and where I now work. Instead, discerning drinkers can choose between domestic and imported bottled water, including Fiji Water, offered in its distinctive rectangular bottles. The name is not a marketing gimmick: the water is bottled in Yaqara, Fiji, 8,100 miles from Cambridge.

How the water got from Fiji to Cambridge illuminates the logistics of our global economy. The Canadian businessman

David Gilmour used the fortune he'd made from a Nevada gold mine to purchase Wakaya, Fiji, an uninhabited 2,200-acre island ringed with white-sand beaches. Originally intending to use the island as a family retreat, Gilmour eventually realized its potential as an exclusive resort. Flown in on Gilmour's six-seat plane, guests paid thousands of dollars a night to stay in thatched-roof villas, eat gourmet food prepared from "native venison, vegetables and herbs," and sip French champagne and Evian water. Gilmour tells reporters that he saw a guest guzzling Evian as he played golf and realized he needed a local alternative.

Gilmour's next steps suggest that his ambitions were a bit larger than providing ecofriendly bottled water to the occupants of his nine villas. In 2003, he leased fifty acres on Vitu Levu, Fiji's largest island, obtained a ninety-nine-year lease on the area's underground aquifer, and invested $48 million in building a state-of-the-art bottling plant. He then hired Doug Carlson, an Aspen, Colorado, hotel executive, to build Fiji Water into a global luxury brand. Carlson introduced the water to the American market through gourmet restaurants, persuading chefs to serve the bottle in a silver sleeve and sell the water for $10 a pop. Early adopters included movie stars and musicians, whose patronage helped turn the bottles into a popular fashion accessory, and an "aspirational brand" affordable to a mainstream audience.

In 2004, Gilmour sold the company to the American entrepreneurs Stewart and Lynda Resnick, who had made their fortune selling collectible knicknacks marketed by the Franklin Mint. The Resnicks promptly rebranded Fiji as a green company, buying carbon offsets for the environmental costs of shipping bottle blanks to China, empty bottles to Fiji, and filled bottles to the United States and beyond. Concerns about the carbon footprint of the product have had little impact on sales; by 2008, Fiji Water had outpaced Evian as the top-selling "premium bottled water" brand in the United States.

The Danish shipping giant Maersk routes shipping containers from Suva, Fiji, to Cambridge, Massachusetts, via Auckland, New Zealand, and Philadelphia's Packer Avenue Container Port. Maersk's online shipping calculator tells us that the voyage takes thirty-three days and that transporting a forty-foot container costs $5,540.30, including trucking the box from Philadelphia to Cambridge. These containers can hold over 30,000 kilograms, which means the shipping cost for a liter of Fiji Water from Suva to Massachusetts is roughly eighteen cents. Atoms may be heavy and hard to move, but transporting them halfway across a planet is shockingly inexpensive.

It's surprisingly easy to move atoms from Fiji to the United States. Moving weightless bits along the same route is more complicated.

Fiji has had a rough history in the years since Gilmour bought his private island. Tensions between Fijians of Melanesian descent and Indo-Fijians led to two coups in 1987, and in the new millennium Commodore Josaia Voreqe Bainimarama, commander of Fiji's military forces, has taken over the government twice, first in 2000 and again in 2006. In 2009, Fiji's supreme court ruled Bainimarama's 2006 coup illegal and demanded he step down. His political allies responded by abolishing the constitution, sacking the judiciary, and replacing the fired judges with rented judges from Sri Lanka. Concerned about negative publicity Bainimarama ordered foreign diplomats and journalists out of the country and instructed remaining reporters to practice "the journalism of hope," and report only positive stories, lest their publications be shut down.

The government of Fiji hasn't needed to threaten US journalists to prevent negative reporting. Commodore Bainimarama's recent address to the UN General Assembly, where he apologized that his country wouldn't be able to hold elections until 2014 at the earliest, received no coverage in New York newspapers. Fiji Water's efforts to go green, on the other hand, merited two features in the *New York Times*.

Fiji Water is apparently more mobile than Fijian news. And it's safe to say that more people have sipped the nation's well-traveled water than have sampled the music of Rosiloa, one of Fiji's leading pop bands, or watched *The Land Has Eyes*, Fiji's first locally produced feature film.[2]

Could it be that atoms are more mobile than bits?

Fiji Water offers a glimpse of a possible future, in which we are free to encounter the best the world has to offer, not just products, but also people and ideas. But our ignorance of Fijian politics and culture suggests that this possible future is far from realized. We need to take a close look at the reality of globalization, not just the promise, to understand the challenge we face: we are increasingly dependent on goods and services from other parts of the world, and less informed about the people and cultures who produce them.

What follows isn't an argument for or against globalization. Instead, it's a portrait of incomplete globalization, a confusing world where some globalist aspirations are realized and many are not. Our challenge is to discover the information and the perspective that allow us to thrive in this incompletely globalized world. A good first step is to establish a map of the territory.

Reconsidering Flatworld

In the last decade we've witnessed the rise of a view that "the world is flat," popularized by the *New York Times* columnist Thomas Friedman in a book of the same name. In a flat world, we are told, communications technologies allow companies to build global supply chains, outsource work, and collaborate across international borders. A US business might manufacture in China, offer customer service in India, and rely on the best minds of Japan and the Netherlands to produce new products, because it's easy to discover the best talent all across the world. As a result,

American workers should think of themselves as competing with
the best and the brightest from every corner of the globe.

Not everyone finds this vision especially accurate. It's certainly
not a new one.

The economist John Maynard Keynes offered a similar view of
a communications-enabled globalization in 1919:

> The inhabitant of London could order by telephone, sipping
> his morning tea in bed, the various products of the whole
> earth—he could at the same time and by the same means
> adventure his wealth in the natural resources and new enter-
> prise of any quarter of the world—he could secure forthwith,
> if he wished, cheap and comfortable means of transit to any
> country or climate without passport or other formality.

What's particularly striking about Keynes's vision is that he
was looking not to the future but to the past. The quotation is
a description of life in London before the First World War closed
an era of rapid globalization.[3]

Two massive wars and a worldwide economic collapse not
only disrupted flows of atoms, people, and bits across national
borders. They also bred strong skeptics of the idea that political
interconnection, through organizations like the United Nations,
would protect sovereign economic interests. In her "maiden
speech" to Congress in 1943, the playwright, journalist, and representative from Connecticut Clare Boothe Luce urged her compatriots not to cede control over international air travel to the
United Kingdom and lambasted Vice President Henry Wallace's
view of an internationalized, interconnected post–World War II
world as "globaloney."[4]

In his 2011 book *World 3.0: Global Prosperity and How to
Achieve It*, the business strategy professor Pankaj Ghemawat
offers a flood of statistics designed to combat contemporary
globaloney. To take one atom-based example, exports represent

about 20 percent of global production as determined by GDP, and Ghemawat maintains that this figure overcounts the real impact of trade. The components that make up your mobile phone are double-counted in international trade statistics: once when they're sold as components and again when they're sold as a finished product. Money prefers to stay at home: venture capitalists invest 80 percent of their capital in home markets, and less than 20 percent of stock market shares are owned by foreign investors. Even apparently interchangeable commodities like rice are surprisingly immobile—only 7 percent of rice is sold across international borders. We are facing a flattening world, Ghemawat concedes, but we should not accept Friedman's vision of a flattened world. Globalization is an incomplete, changeable process that's still unfolding.

The flat-world view looks at infrastructures of connectivity and conflates what *could* be with what *will* be. It blurs three separate phenomena—the globalization of atoms, people, and bits—into a single trend. The infrastructures that it celebrates—container shipping, air travel, and the Internet—quite obviously have the potential to shrink distance and integrate economies and cultures. But they are held in check by social, legal, economic, and cultural forces that make the blurring of international borders a slow, gradual, and uneven process. Understanding the current balance between forces that connect us and those that disconnect us allows us to consider our blind spots and determine whether we're getting what we want and need from the wider world.

In this chapter, we look first at the globalization of atoms to understand how the visibility of internationally mobile bits can blind us to the locality of most of the objects we encounter. Next, we look at human migration, where even small amounts of mobility often lead to fierce political debates. Our tendency to overestimate the movement of atoms and people should give us pause when we think about the world of information, as we consider the idea that bits can be less mobile than atoms.

When we focus on the infrastructures that make globalization possible—the ports and routes of container shipping, the hubs and spokes of air travel, the routers and cables of the Internet—it's easy to imagine a level of connection that's higher than what we currently experience. Understanding how atoms, people, and bits flow—and don't flow—through the world requires less speculation and more observation, including close observation of how bits flow through our computers and our minds.

How Flat Are Atoms?

The Monday after Christmas 2004, Sara Bongiorni, a business reporter for the Baton Rouge *Advocate*, decided that her family would spend the following year boycotting China. The trigger for her decision was the discovery that twenty-five of the family's gifts under the Christmas tree came from China, and fourteen from the rest of the world. Her book documenting their experiences, *A Year without "Made in China,"* chronicles the small crises a China boycott creates for Bongiorni and her family: shoes for her toddler made in Texas cost seventy dollars rather than ten dollars for shoes made in China; buying an inflatable pool for her kids became impossible without violating the boycott.

Had Bongiorni's rule set required her family to forswear products with *components* made in China, the task would have bordered on the impossible. Today's manufacturing supply chains span the globe, and apparently simple products include inputs from many nations. As a student at MIT's Media Lab, Leonardo Bonanni developed a platform called Sourcemap that helps customers and companies document the global origins of everyday objects. A pair of denim jeans "made in Indonesia" includes cotton grown in the United States, processed in China, woven in Thailand, cut in Singapore, and sewn together in Indonesia, using thread from Malaysia, rivets from Taiwan, and a zipper from Hong Kong.[5] Unpacking the origins of these simple objects reveals the power and pervasiveness of global sourcing, low-cost shipping, and "just in time" inventory systems.

Six months into her experiment, Bongiorni visited Walmart to test a claim made by Mona Williams, Walmart's VP of corporate communications. Williams, in a letter to *Newsweek* magazine, asserted that Walmart purchases vastly more from US suppliers than from Chinese suppliers. Bongiorni, wandering the aisles of Walmart, found the numbers hard to believe. After all, Walmart sources as much as 70 percent of its nonfood inventory in US stores from over five thousand Chinese suppliers, which makes the American retailer China's eighth-largest trading partner, ahead of Russia, Australia, and Canada. Bongiorni spent an afternoon in Walmart, checking 106 items for their country of origin, and found 49 percent made in China and 22 percent made in the United States, with Honduras running a distant third. "The way I see it, unless Ms. Williams is including groceries, or building supplies used to construct Wal-Mart stores in her tally of American purchases, I just don't see how her numbers add up," Bongiorni wrote.

Groceries are a good place to start unpacking Bongiorni's misperceptions of the global economy. Walmart is not just the largest retailer in the world; it's also the largest grocery store chain in the United States. And despite eye-catching examples of globalization like water shipped from Fiji or lamb chops from New Zealand, less than 7 percent of the food consumed in the United States comes from outside our borders.[6] Since 54 percent of Walmart's sales come from groceries in 2011, that's a lot of American-made foodstuffs that Bongiorni didn't consider.[7]

Her instinct that the materials used to construct Walmart stores are sourced locally is a sound one as well. While it's inexpensive enough to ship high-value goods like electronics and, remarkably, drinking water, and still make a profit, building materials are another matter. Steel, timber, and concrete used in the United States are primarily sourced domestically—about 20–25 percent of the steel used in the United States is imported, as is under a third of the construction timber, imported mainly from Canada.[8] The company's relationship with suppliers of building materials, construction companies that build Walmart stores, refiners that supply gasoline and diesel fuel to Walmart trucks, and contractors that clean the stores are less visible than the "made in China" labels that so haunt Bongiorni.

Bongiorni's visceral reactions to the artifacts of globalization helps us understand how we can overestimate the globalization of atoms. She is particularly incensed by items she considers specifically non-Chinese that are manufactured there: a ceramic statue of Jesus, patriotic decorations for the Fourth of July. These distinctly American items, she feels, should be made in America, and their Chinese origins are proof to her of America's manufacturing decline and China's rise.

The Chinese-made ceramic Jesus and the bottle of Fiji Water both invite us to imagine a level of globalization that's higher than actually exists. The French economist Daniel Cohen observes, "French people 'see' a McDonald's on every corner, American films in all the theaters, Coca-Cola in all the cafeterias, but they are apparently blind to the thousands of French cafés that serve ham-and-butter sandwiches, the bottles of Evian and Badoit, the French films featuring Gérald Depardieu, or the regional press. In wealthy countries, globalization is largely imaginary."[9]

Geography still matters.[10] Despite the rising globalization of atoms, we have a strong bias in favor of local products. In 2000, the economist Jeffrey Frankel calculated a theoretical level of globalization with which we might compare our actual levels of global trade. The United States represents roughly one-fourth of the world's economy. In a truly borderless world, we would expect Americans to buy and sell 75 percent of their goods abroad. In fact, America's international purchases and sales equal roughly 12 percent of GDP, suggesting a level of globalization that's roughly one-sixth what we might anticipate in a flat world where national origins of a product had become completely irrelevant. And while China is the United States' second-largest trading partner (neighbors Canada and Mexico are first and third),[11] products made in China represent only 2.7 percent of US consumer spending.[12]

One reason that atoms are still so immobile is that governments excel at slowing their flow. Protecting domestic markets is so tempting for governments that they often hinder the flow of global atoms with one set of regulations, while promoting free trade with other legislation. The United States is particularly practiced at this form of economic hypocrisy when the atoms in question are grown by farmers.

In theory, one benefit of globalization is the ability for economies to specialize and take advantage of the specific strengths of their workforce. Wealthy countries tend to have well-educated and expensive workers. It makes sense for them to design and manufacture high-value, technically complex goods like computers, electronics, and machine tools. Poor nations have undereducated and inexpensive workers, whom we would expect to be employed in sectors like agriculture and mining, producing raw materials at low cost for export to wealthier nations. In this paradigm Mali grows cotton and exports it to China, where it's woven and sewn according to Italian designs for export to the United States.

Not quite. As it turns out, the United States is the world's largest cotton exporter, responsible for roughly 40 percent of the world's cross-border cotton trade. We would expect a wealthy nation with high labor costs to leave production of an agricultural commodity to developing nations, but America's dominance in this market is made possible by massive agricultural subsidies that started in the 1930s. These subsidies averaged $3 billion a year over the past decade and have ensured that the 25,000 American farmers who grow cotton receive roughly twice the market price for their goods.

Because American farmers have such great financial incentives to grow cotton, they grow a lot of it. The United States trails only China and India in total cotton production. And American farmers can sell cotton very cheaply, depressing global markets, because the price they are paid is set by the government, not by the market. Brazil, a major cotton producer, was so incensed by the US system that it sued the United States through the World Trade Organization and won a settlement in which the United States pays Brazil $147 million a year for the right to continue subsidizing domestic cotton production.[13]

The Lumpy World of Migration

Economic logic suggests we should see a world where atoms are profoundly mobile. Instead, cultural preferences and government regulation shape a world of atoms that's more local than global, though that reality can be difficult to see at first. But while the artifacts of globalization can blind us to the more complex realities on the ground, the fiercest emotions about globalization tend to be reserved for discussions of the movements of human beings. If we have a hard time seeing the incomplete globalization of atoms, understanding the realities of human mobility is even harder. What's emerging is not a single trend but a complex pattern, hidden behind a shrill, popular narrative that sees immigration as a uniquely modern crisis. When we see only that oversimplified narrative, we miss the simple fact that developed nations need more migration, not less.

Hostility to migration is becoming mainstream in some European nations. Anti-immigrant political parties like France's National Front and Greece's Golden Dawn are emerging as influential political actors and members of coalition governments. Other European nations are reconsidering liberal immigration policies in the face of rising populations of Muslim immigrants who, Europeans fear, won't integrate into society as completely as previous immigrants have. In the United States, a sustained recession has led some unemployed people to speculate that their joblessness is caused by illegal migration. Campaigns to ban the burka in France and to enshrine English as the official language in United States may suggest that people are as uncomfortable with images and speech that remind them of migration as they are with migrants themselves.

Support for these and other initiatives, as well as the popularity of anti-immigrant politicians, might imply that countries are experiencing unprecedented levels of immigration. In fact, global migration is significantly lower than it was a hundred years ago. Just before World War I, roughly 10 percent of people worldwide lived in countries other than the land of their birth.[14] Mass migrations from Italy, Ireland, Norway, and Germany reshaped the United States, Canada, and Argentina, shifting 27 million Europeans overseas in the three decades before World War I. Prior to these voluntary migrations, indentured servitude sent Chinese and Indian workers to Africa and the Caribbean and Africans to North and South America. Today, immigration seems high to many of us only because global mobility slowed almost to a stop after World War II and remains well below historical highs.

A German farmer leaving for Minnesota in 1910 faced a long and dangerous voyage, an uncertain and risky future, hostility and resentment from his new neighbors, and a near-complete severing of existing social ties. Technological change in the intervening decades offers a contemporary migrant a very different prospect. Jet travel means a (legal) voyage is essentially riskless and instantaneous in comparison with overland and ocean journeys. A Nigerian immigrant to Houston can call home for a few cents a minute or via Skype for virtually no cost. She can read Lagos newspapers online and download the latest Nollywood films. It's comparatively inexpensive to return home for a visit or to migrate in reverse. Some social theorists have begun to worry that it may be possible to migrate physically but not culturally, offering examples of Pakistani and Turkish communities in northern Europe where Urdu and Turkish remain the dominant languages and residents encounter more satellite television than local media.[15] This phenomenon—physical mobility with cultural stasis—gives fuel to politicians who argue for banning the burka or for English-only education.

While it's possible to stay more closely connected to home than ever before, these technological developments haven't radically increased the volume of international migration. The International Organization for Migration estimates 214 million migrants worldwide, or 3.1 percent of the world's population. That percentage is rising from a post–World War II nadir, but slowly. Between 2000 and 2010, the migration rate increased from 2.9 percent to 3.1 percent of the global population.[16] Far, far more people want to migrate than are able to, but they are constrained by immigration restrictions.

The rise of outsourcing can be thought of as one response to a world where jobs are mobile but people are not. Many of the Indian employees who answer customer service phone calls from Bangalore would be interested in living and working in Europe or the United States. They're able to work at a distance because the forces that constrain their movements don't apply to the bits traveling between computers and across phone lines. If migration restrictions were relaxed or eliminated, it's likely that millions of people would move to places where they thought economic and political conditions were better, enabled by globalizing technologies like air travel and inexpensive telecommunications. But this possible movement is held in check by laws that seek to protect cultures, economies, and social welfare systems from too much flattening.

The result is a world with deeply uneven migration patterns. In some nations that are highly dependent on "guest workers," migrants who reside in the country for extended periods but don't have the rights of citizens, immigrants represent the majority of the population: 87 percent in Qatar, 70 percent in the United Arab Emirates, 69 percent in Kuwait. Other nations have little to no immigration, because of a lack of economic opportunity (Indonesia, 0.1 percent; Romania, 0.6 percent) or legal or cultural barriers to migration (Japan, 1.7 percent; South Africa, 3.7 percent). Levels of immigration in North America and western Europe are higher than the global average, which makes sense, because migrants generally leave poorer nations for wealthier ones. Some 9.39 percent of the population of the European Union live in countries other than the country of their birth, as does 13.9 percent of the US population and 21.3 percent of those living in Canada.[17]

There's no clear threshold at which migration triggers societal tensions and debate. South Africa has seen anti-immigrant violence even though more than 96 percent of its population is native born, while Canada promotes a multicultural identity as part of its strength as immigrants now account for more than 21 percent of the population. What may be more important than an absolute number or percentage of migrants are perceptions of how migration is changing a society.

Rhetoric around immigration in Europe includes the idea of a "time bomb" of Muslim migration, often accompanied by projections that Muslims will represent 20 percent of the population of the EU by 2050.[18] Those projections are extrapolations from existing migration patterns and birthrates. The scenario in which one of five Europeans is Muslim is the "naïve" scenario postulated in a controversial article by a little-known Hungarian academic; more sophisticated models (which assume that birthrates for Muslim families in Europe will drop as those families become better educated, a well-documented demographic phenomenon) show an even smaller growth in the Muslim population.[19]

It may also be worth noting that Islam has been one of the world's fastest-growing religions over the past half century. The

Pew Research Center's Forum on Religion and Public Life projects that 26.4 percent of the global population will be Muslim by 2030, suggesting that even if the prediction of the European "time bomb" proves to be accurate, Europe would have a much lower proportion of Muslims than the world as a whole. Pew predicts that 8 percent of Europe will be Muslim by 2030, and that the United States will have a larger Muslim population than any European nation other than Russia or France.[20]

Rates of migration are much higher than they were forty years ago, as we've moved beyond postwar isolation toward a rate of migration closer to what the world experienced in 1900. And migration undoubtedly raises challenges for governments and societies. But it's worth keeping in mind that the illusions we hold about the mobility of people can blind us to the actual demographic challenges nations face. One of the reasons European nations are loath to impose sharp restrictions on immigration is that their populations are aging. Without an influx of young taxpayers, they run the risk that their social welfare systems won't be able to support a population of elderly retirees. Focusing on the illusion of a Muslim takeover of Europe, or even the illusion of a flat world where labor is highly mobile, makes it very difficult to see and address problems like maintaining a survivable worker-to-retiree ratio. In a flat world, Indians now at work in call centers might be flocking to Japan to care for the elderly. In our semiglobalized world, they're held in place by immigration restrictions and cultural barriers.

Bits: Potentially Mobile, Practically Static

We see a "made in China" label and imagine the end of American manufacturing. We see minarets and imagine a wave of Muslim immigrants to Europe. But the fantasies of a seamlessly globalized world of bits are even more seductive, since the bits are produced and marketed by the world's biggest technology companies.

It's dark in the conference room, where stern-faced Japanese businessmen confront an intractable problem: their sole supplier wants too much money for valves! One of the youngest executives, the only man seated in front of a computer, announces that they've gotten an online bid from "Mitchco" for half the price. "Where are they?" asks the boss. The young executive answers, "Texas," and we cut to a dusty machine shop, where Mitch, in his "Mitch & Co." coveralls, looks at a screen and announces in a drawl, "Domo arigato."

The ad, aired in 2000, advertised IBM's eCommerce division, which offered "solutions for a small planet." That small planet is connected by more than the planes and container ships that deliver Mitch's valves to the Japanese factories; it's made possible by the flow of bits that allow Mitch to learn about the opportunity in Japan and bid on the contract.

The international flow of bits suggests a cosmopolitan future, where Texas-based mechanics learn Japanese to conduct business with their new partners. But as with the promises of globalized atoms and people, it's worth looking closely at the gaps between the potential and the actual globalization of bits. If technological advances have increased the potential mobility of atoms and people, they have utterly transformed the potential mobility of bits. An international phone call cost $2.43 a minute in 1970. As powerful fiberoptic networks replaced overloaded copper ones and rival providers offered competitive services, that cost fell to $0.14 a minute in 2004, and the volume of international phone traffic rose from 100 million minutes in 1970 to 63.6 billion in 2004.[21] More recently, the rise of the mobile phone has put the possibility of international connection into almost everyone's hands. In 2000, roughly 740 million people owned mobile phones. By 2011, there were nearly 6 billion mobile phone subscribers, or 85 phones for every 100 people on the planet.[22]

Technology hasn't just made communications cheaper. It has made the previously impossible seem routine. Once a month, I chat with a friend in Budapest over Skype. Using the video cameras embedded in our laptop computers, we show off our children to each other and they babble to each other, separated by five thousand miles but connected by Internet service that's too cheap to meter. As recently as ten years ago, this sort of ad hoc, home-based international videoconferencing wasn't just prohibitively expensive: it wasn't possible.

That's the appeal of the Mitchco IBM ad. The impossible—our friend Mitch acting as supplier to a major Japanese auto firm—is made real by IBM's miraculous technology. In a world where business relationships last longer than the sixty seconds of a television commercial, Mitch may want to start reading up on his new client and its competitors. Again, technological progress has changed the landscape and rendered the impossible routine.

Twenty years ago, the average citizen had access to a few television channels, each of which was broadcast from a transmitter within a few hundred kilometers. Now, with only a small satellite dish, an interested viewer in Ghana can tune into programming from dozens of nations. An aficionado of international news twenty years ago frequented newsstands in major cities, waiting to buy stale copies of *Le Monde* or the *Times of India*. Those newspapers are now available online, instantaneously, for anyone with Internet access and the inclination to peruse them. And most of them are free. Newspapermap.com offers links to more than ten thousand newspapers from more than one hundred countries, accessible online and machine translated from more than a dozen languages. Moreover, access to information online goes well beyond newspapers. As thousands of Americans discovered during the January 2011 Tahrir Square protests, Al Jazeera English is available online, as are streaming video services from Japan's NHKWorld, France24, Russia's RT, and dozens of others.[23]

And yet, we rarely encounter this bounty of international information. If the global flow of atoms is constrained by trade restrictions and by taste, and the flow of people by employment opportunities and immigration laws, the flow of information is constrained by our interest and attention. Much as we imagine a flat world of migration when visiting Dubai or a flattened supply chain when in the aisles of Walmart, looking at the richness of news available across international borders conjures up a world where we're deeply connected to information and perspectives from around the world. The reality is more complicated.

The *Times of India* has a print run of 3.1 million copies, giving it the largest circulation of any English-language daily newspaper in the world. The online edition of the paper has an audience of roughly 9.1 million users in an average month; 1.1 million of those users are in the United States. It's fair to assume that the advent of online newspapers has dramatically increased the US readership of the *Times of India* from the days when reading the paper required it to be flown in from Mumbai.

Americans generate about 800 million page views a month to large international news websites like the *Times of India*, and more than 10 billion to domestic news sites. In a theoretical flat world with attention distributed equally among all corners of the Internet, like the one Jeffrey Frankel proposes we consider in terms of trade, Americans, who represent only 11.2 percent of the Internet's users, would get 88.8 percent of their news from other countries. In practice, we get a lot less.

Doubleclick, an online ad company that has become part of Google, publishes statistics on the monthly web traffic and "reach"—the percentage of Internet users in a country who visit a particular site in a month—for tens of thousands of news sites. Because Doubleclick analyzes traffic in more than sixty countries, and because it categorizes sites by topic, we can ask which one hundred news sites are most visited by Americans or South Koreans, and whether those sites are domestic or international.

In the view from the United States, the BBC is the eighth–most popular news site, and the UK papers the *Guardian*, the *Telegraph*, the *Daily Mail*, the *Times*, and the *Sun* all have substantial US audiences. The *Times of India* ranks ninety-fourth in popularity with US news readers, and it is the first non-UK, non-US source on the list of news sites popular in the United States. Of the 9.87 billion page views Americans generated to those top hundred news sites in July 2010, 93.4 percent were to sites hosted in the United States, while 6.6 percent were to international sites like the BBC and the *Times of India*.

It's possible to conclude that Americans are parochial, and as such less likely to read international news sources than their cosmopolitan brethren in, say, France. But look at the numbers: 98 percent of the traffic to the top fifty news websites in France goes to domestic sites, 2 percent to international sites. In China, the first international site to appear on a list of top news sites is Reuters.com, in 62nd place, followed by the *Wall Street Journal*, in 75th, and the BBC, at 100th.[24] Of the ten nations with the largest online populations that Doubleclick checks, the United States ranks as one of the least parochial nations, which is likely due in part to our large student and immigrant populations. None of the top ten nations looks at more than 7 percent international content in its fifty most popular news sites.[25]

It makes sense that linguistically isolated nations—nations that don't share a principal language with any other countries—like Japan or South Korea would read few international sources. It's more surprising that despite a long colonial legacy, and shared language, Indians and Brits don't read more of each other's content. Nor do they visit US news sites. As it turns out, shared language offers no guarantee of interest in each other's media. Spanish-speaking nations in South America show little or no interest in reading each other's news or in online news from Spain. There is often interest in news across borders from smaller nations that share a language and a border with a larger nation:

Percent of news page views from domestic sites,
Google Ad Planner data, 6/2010.

Internet users in Taiwan and Hong Kong read a lot of mainland Chinese news, and Canadians read American news more avidly than vice versa. (The Canadians, Hong Kong Chinese, and Taiwanese may all be keeping an eye on their powerful and unpredictable neighbors.) Interest in specific topics, especially technical topics, seems to draw people across online borders. Aside from the BBC, which shows up in virtually every country's media profile, tech-focused websites like C|Net are the most likely to attract international readers.

So who are the 1.1 million Americans reading the *Times of India*? Are they entrepreneurs, like Mitch, looking to understand the trends in promising new markets? They're an advertiser's dream: the majority report an annual income of over $75,000, and 70 percent have either a bachelor's or a graduate degree, which means they are wealthier and better educated than the online audience of the *New York Times*. They're also very loyal. In a given month, they generate 60 million page views and visit the site eleven times a month.

While some may be curious entrepreneurs, the vast majority of these avid readers are members of the 2.8 million strong community of "nonresident Indians," a term the Indian government uses to include Indians living in the United States on short-term visas as well as those who've become US citizens.

The *Times of India* data suggest that it's too simple to say Americans get 6.6 percent of their news from international news sites. Some Americans, like nonresident Indians living in America, get much of their news internationally; these readers make the rest of us look more cosmopolitan than we actually are. It's not surprising that the Internet hasn't magically caused most Americans to get their daily news from the *Times of India*, or that Indian Americans are disproportionately interested in Indian news. We pay attention to what we care about and, especially, to persons we care about. Information may flow globally, but our attention tends to be highly local and highly tribal; we care more deeply about those with whom we share a group identity and much less about a distant "other."

If the flow of bits is constrained by interest and attention, it raises an uncomfortable question: Are we getting enough information about the rest of the world in order to flourish in a world of increasing connection? We need this information to thrive in a connected world, whether our goal is landing international business contracts or responding to threats like SARS.

Perhaps Mitch hasn't yet discovered Asahi Shimbun's "Asia and Japan Watch,"[26] with regular English-language coverage of Japanese economics and "cool Japan." Maybe he's counting on the *Houston Chronicle* to bring him Asian news from a US perspective. That would be a bad choice on his part.

American Journalism Review has conducted a "census" of foreign correspondents writing for US newspapers since 1998. Since their study began, twenty US newspapers have cut their foreign bureaus entirely, and the 307 correspondents *AJR* was tracking

in 2003 had shrunk to 234 by 2011.[27] Fewer dedicated correspondents doesn't necessarily mean less international coverage in US
newspapers; newspapers are leaning more on "parachute correspondents" and on news wires to cover international stories.
But the number of stories is dropping as well. The Project for
Excellence in Journalism surveyed sixteen newspapers between
1977 and 2004 and saw a drop in front-page coverage of "foreign
affairs" from 27 percent of all stories to 14 percent.[28] My team at
MIT's Center for Civic Media conducted a similar study, looking at all the stories published in four major US newspapers in
four weeks evenly spaced between 1979 and 2009; we saw only
one-third as many international and foreign stories in two of four
papers, and a significant drop in the third. (We found no significant drop in the *New York Times* over the same time period.)[29]

Television news in the United States has seen a similar, dramatic drop in international coverage. Some 78 percent of Americans report getting news from local television stations, and 73
percent from national broadcast or cable new coverage. A study
of American television news from Harvard's Shorenstein Center
reports that 45 percent of American television news stories in
the mid-1970s delivered international news.[30] Working with data
from the Vanderbilt Television Archive and the Project for Excellence in Journalism, Alisa Miller, a journalism scholar and president of Public Radio International, estimates that 10 percent of
recent stories on national news broadcasts and 4 percent of local
news broadcasts were international news stories.[31]

Despite the sharp fall in the supply of international news in
the United States, audiences don't seem especially concerned. A
survey by the Pew Internet and American Life Project found that
63 percent of Americans believe they're getting sufficient international news, while only 32 percent see a need for more coverage.[32]
Respondents want more state and local news, and more stories
on religion, spirituality, and scientific discovery. Less than 40 per-

cent of Americans follow international news closely,[33] which may help explain why there is less space for international stories in newspapers and television. Some percentage of Americans are seeking out international news via public radio and via the Internet, though our data on visits to international news sites suggest that the population motivated to seek out different stories and perspectives is quite small.

Of course, Mitch might learn about his new Japanese clients through other means. Through Netflix, he might rent some Kurosawa films after a long day in the machine shop. Again, there's a gap between opportunity and practice. Netflix reports that interest in non-US film has remained low throughout its corporate history, representing 5.3 percent of rentals in 1999 and 5.8 percent in 2006.[34] And if Mitch visits a bookstore to look for a Haruki Murakami novel, he'll discover that only 3 percent of the books published in the United States are works in translation. (The numbers are even lower for fiction and poetry—usually lower than 1 percent.)[35]

IBM's Mitchco ad asks us to imagine a connected future. The actual appetite for news and film that cross international borders suggests that this connected future may be a fantasy. If atoms and people are prevented from crossing borders by tariffs and laws, bits are slowed by our interests and preferences, which are probably even harder to change than government trade policies.

This vision of a globally connected, informed, and cosmopolitan world isn't just the product of a single IBM ad. It's part of a narrative offered by the individuals and companies building the Internet. This narrative is both a marketing campaign and an inevitable consequence of our imagination. Powerful new infrastructures invite us to imagine profound changes. To understand what the Internet is and isn't doing, we need to look at the network from at least two different angles. We have to look at what's possible and what actually happens, at a map of infrastructure and a map of flow.

Infrastructure and Flow

There are at least two ways to draw a map of San Francisco. You can start from satellite photographs, tracing the routes of streets, the coastline, and the locations of key buildings. The artist Amy Balkin chose to make a radically different sort of map. "In Transit" is drawn by means of data collected from thousands of Yellow Cabs as they travel the streets of the city. San Francisco's Yellow Cab company uses GPS to track the location of its cabs, and the company released a large set of this data—stripped of information that would identify drivers or passengers—to a set of graphic designers, who built portraits of the city from this information.[36]

Balkin's map reveals the major highways and streets of the city as thick, white lines of light, the aggregated path of hundreds of taxi journeys. The city's coastline and major parks emerge as dark spots where taxicabs can't go. Other dark spots reveal neighborhoods where taxis rarely go, like Hunter's Point, a historically African American neighborhood in the south of the city.

A map of flow is likely to be less complete than a street map of a city, but it conveys information the traditional map lacks. On Balkin's map, it's easy to see paths from area airports to San Francisco's downtown, and east–west paths from the downtown to Pier 39 and other waterfront tourist attractions. But one can also see a set of north–south paths, drawn by cabs acting as ad hoc ambulances, that link residential neighborhoods and hospitals. The ley lines of the city become visible.

You might not choose to use Balkin's map to navigate from Union Square to Fisherman's Wharf, but its the map you'd want if you were a city planner considering new bus routes or an entrepreneur looking for a busy corner on which to site a gas station. Traditional maps of infrastructure show you all possible paths people could take, whereas flow maps show you the paths people

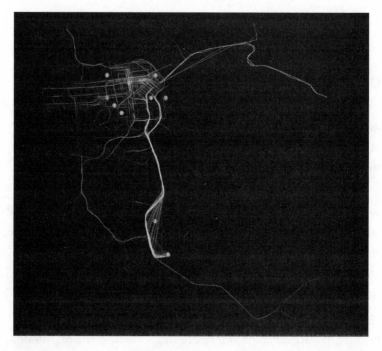

San Francisco taxi routes. From "In Transit," by Amy Balkin, 2006.

actually choose. And if most tourists walk to Fisherman's Wharf on Stockton Street, knowing that Stockton and Beach is a busier corner than Taylor and Beach could be the difference between success and failure of your new T-shirt store.

Many of the maps we encounter in our lives map infrastructures. A road map shows us where we're able to drive in a car; a transit map shows us places we can access easily by a train, subway, or bus. Cellphone coverage maps show us where our mobile phones will (and will not) work. Maps of this sort are undeniably useful, but they can also be deceptive. Knowing that we can get from here to there doesn't tell us whether the route is so popular that it's likely to be jammed with traffic, or so unpopular that it may be circuitous, dangerous, or hard to follow.

The history of industrialization is told in part by maps of infrastructure. Railroad companies were the first to mass-produce maps, working with printers to perfect lithographic techniques

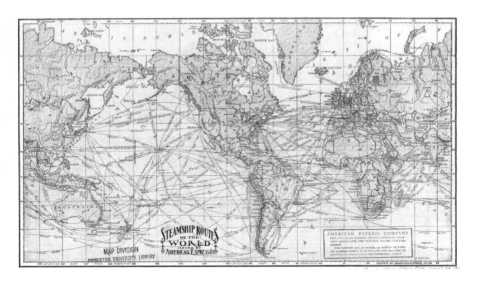

American Express, Steamship Routes of the World, circa 1900.

and produce intricate, detailed maps of railway lines running deep into the American frontier, or knitting together Britain's factories, mills, and ports. In nineteenth-century America, the maps were literally propaganda. For railroad companies to turn a profit, they needed to sell the land bordering the tracks, granted to them by an act of Congress. Maps were printed in the languages spoken by new immigrants to East Coast cities to encourage them to spread inland. The least honest of these maps featured cities with familiar names—Crete, Dorchester, Exeter, Fairmont—neatly lined up in alphabetical order along a rail line. That these cities had yet to be built didn't decrease their appeal to newly arrived immigrants; the maps showed an empty wilderness as civilized and connected, with the railroad as land broker, conveyance for plows, seeds, and other supplies, and the only way back to the communities they would leave behind.[37]

Almost as intricate as a Rand McNally atlas of railroad lines are contemporaneous maps of telegraph lines, showing the town-to-town connections that made it possible to send messages from Louisiana to Nova Scotia. Or the American Express Company's

1900 map of steamship lines, where the oceans disappear under thick red lines of connections between ports. The message is simple: our infrastructure connects the world, and if you join us, you'll be connected too. These maps don't help you navigate; they don't help you drive the train or steer the ship. They're maps of the possible.

The wave of maps that accompanied the rise of the commercial Internet echoed this age-old message. Network providers offered maps of optical fibers connecting major cities that looked hauntingly like early railroad maps. (In many countries, fiber-optic cables followed rail lines, so early Internet maps were essentially rail maps.) The geographer Martin Dodge collected hundreds of early maps of the Internet, including maps of physical networks as well as maps of "topologies," the ways in which these networks routed traffic to one another.

As networks became more pervasive and complex, these topologies took on an organic quality. The Opte project, completed in 2005, was one of the last efforts to visualize the topology of the connected Internet. The visualization it produced is a set of multi-colored branches of unfathomable complexity, looking more like images of human neurons than railroad tracks. In fact, it's impossible to read the Opte "map" as anything other than an image, a symbol of an Internet so complex that it should be thought of in organic terms, as part of the natural order of things.[38]

If infrastructure maps hold out the possibility of connection, flow maps offer different promises and different challenges. For one thing, they're harder to make than infrastructure maps. Infrastructure tends to stay still. That's not true with maps of traffic congestion, which can change minute to minute and are invariably different on weekdays and weekends. How can we measure it? For large streets and highways, information is available from state and national departments of transportation, which embed sensors in major roads to monitor traffic volume and speed.[39] Google uses this data to render maps of San Francisco that include close to real-time information about traffic conditions on many of the streets shown on Amy Balkin's cab map. For traffic data on smaller streets, Google asks its users for help. When you use Google Maps on your phone, you send information on your position and speed to Google's servers, which aggregate this data to make informed guesses about the speed of traffic on the street you're using.[40]

In a blog post alerting users that their data was being harvested to make traffic maps, Dave Barth, the product manager of Google Maps, referred to this practice as "the bright side of sitting in traffic." Aware that some users might consider this "bright side" an invasion of privacy, Barth assured users that the data Google uses to make maps is anonymized, and that they could opt out of data collection. This reasonable privacy concern highlights a core tension inherent in monitoring the flow of people: the maps produced can be tremendously useful, but making them rapidly veers into the realm of surveillance.

In 2010, the German Green Party politician Malte Spitz sued his mobile phone provider, Deutsche Telekom, for access to all information the company had about his phone usage, as a way of highlighting corporate surveillance. He won the suit, and DT presented him with a spreadsheet containing 35,831 rows of data collected over the six months between August 2009 and February 2010. The information included whom he called and texted and when he checked his email. A clear portrait emerged of when he was awake, asleep, working, or playing, and exactly who was in his social network.

Each of the 35,831 rows of data included geographic coordinates for Spitz and his phone. Mobile phone operators are able to make very accurate estimates of a user's geographic position by measuring the strength of the signal the phone receives from nearby cell towers. Phone operators use this information to provide location information to the police or ambulance services if a user calls an emergency hotline.[41] Spitz worked with the German newspaper *Die Zeit*, which used the mobile phone data and publicly accessible data, like his Twitter feed, to build a map that shows his movements and activities over the period of six months.[42]

Watch the map like a movie, and Spitz roams his West Berlin neighborhood, orbiting Rosenthaler Platz. Focus the map on Nürnberg, and the timeline tells you that Spitz was in town the morning of September 9, 2009, and passed through briefly on November 20, 2009. Zoom in still further, and you can discern the paths Spitz takes through his neighborhood and the beer gardens he favors.

An interviewer asked Spitz what he'd learned by looking at the visualization. He said that he was surprised to discover how small his personal orbit actually was: "I really spend most of the time in my own neighborhood, which was quite funny for me. . . . I am not really walking that much around."[43]

It's easy for Spitz to imagine himself as more mobile and less predictable than he actually is. The major events in his recent past—a speech at a conference across the country, a visit to his hometown—register as signal against the noise of countless trips to the local coffee shop. This cognitive bias is a form of the regression fallacy, a tendency to pay more attention to unusual moments in our lives than to the ones closer to our average, everyday existence.

If we zoom out from Spitz's experience and look at our own movements, we are likely to see our own local biases, the familiarity of the paths we trod. Much as we can imagine Chinese products in every store, shelved by Mexican immigrants, we can imagine ourselves encountering much more of the world than we actually see. Looking at maps of the world, and the maps of our path through the world, helps illustrate the disparity.

In early 2009, the Canadian photographer John O'Sullivan used data from hundreds of airline route maps to produce a vast visualization of those routes. In his image (below), every commercial airline route he could find is represented by an arc between two cities. Routes flown by multiple airlines have thicker arcs,

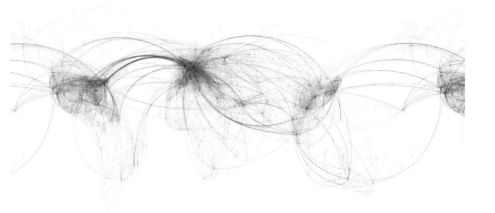

Global airline routes. Map by John O'Sullivan.

which darkens the best-connected cities into key points on the map. From these thin blue arcs, the recognizable outlines of continents emerge—South America tethered to Spain and Portugal, Africa to Britain. O'Sullivan's map shows what's possible with a passport and an infinite supply of frequent-flier miles: access to virtually any spot on the globe.

Dr. Karl Rege and his team at the Zurich School of Applied Sciences used similar data but added another dimension—time—to create a very different map. With data from FlightStats.com, a site that tracks commercial aviation, Rege's team created a video,[44] which renders each airplane as a tiny yellow dot moving across the surface of the earth. The 72-second clip shows a day's worth of flights and reveals patterns that aren't visible from a static map of airline routes. A flock of planes leaves the eastern United States for Europe as night falls in New York, and a reverse flow from Europe to the United States occurs as midday reaches London; a dense net of flights ping-pongs between eastern China, South Korea, and Japan regardless of the time of day; the United Arab Emirates emerges as a midpoint for flights between Europe and Australia; from south to south, there are vanishingly few flights connecting South America, Africa, or Australia.

The most striking pattern in the Rege animation is the thick, yellow mass that covers the United States, Japan, eastern China, and Europe at the peak of their workdays. Even with each flight represented by a single pixel, there's enough commercial air traffic over the United States—almost 25,000 flights per day—to render any features on the ground invisible. There's also vastly more domestic air traffic than international traffic. In 2009, about 663 million passengers departed from US airports. Only 62.3 million disembarked in other countries, and nearly 19 million of those landed in either Mexico or Canada. (And only 32.8 million of the passengers are Americans—the other 29.5 million are nationals of other countries who've traveled to the United States for business or vacation.)[45]

International travel accounts for just 9.4 percent of air passengers and a smaller percentage of commercial flights, as international flights, on average, carry more passengers than domestic ones. The typical traveler at an American airport isn't leaving the country or even headed to the other coast—she's traveling less than 900 miles, visiting a nearby city, often in the same time zone.[46] Rege's visualization suggests that this pattern holds true in other parts of the world as well: Europeans are traveling mostly in Europe, Chinese in China, Japanese in Japan. The infrastructure of air travel is global, but the flow is local.

Imagine a vastly more complicated version of Rege's animation that includes data like Malte Spitz's, but for each person in the world: a Marauder's Map that treats the whole world as Hogwarts. Everyone's daily movements—the train to work, the car trip to the grocery store, walks to the park or playground—would be represented. If we could overlay the trillions of trips people make on foot or by bicycle, bus, and car, the flights on Rege's animation would disappear in a blur of local movement. The sort of transnational travel depicted on O'Sullivan's airline route map becomes a rounding error.

Graph our travels—individually or as a nation—in terms of

frequency and length, and the curve that results shows a "long-tail" distribution: the head shows lots of frequent, short trips, while the long, thin tail shows our occasional lengthy trips. The latter trips may be the ones we remember, but we spend most of our time taking shorter journeys, staying close to home.

When we encounter content on the Internet, physical distance is largely irrelevant; we seldom know whether we're reading a web page hosted nearby or halfway around the world. But we need to consider another sort of distance, a distance between the familiar and the unfamiliar. We celebrate the Internet's ability to put unfamiliar and unexpected content at our fingertips, but we have to be cognizant of the difference between infrastructure and flow.

If we monitor our behavior, our flow across the global Internet, we're likely to find that our online travels resemble our off-line ones. Our interaction with people and ideas from far-flung corners of the world are infrequent, if memorable, and the majority of our interactions are with a small set of people, often people with whom we have a great deal in common.

If we look beyond the infrastructures that make global connection possible, at the flow of our attention on the Internet, we are likely to discover the powerful influence on our behavior exerted by homophily.

Homophily

In the early 1950s, the sociologist Robert Merton began an in-depth study of friendships in two housing projects, one in New Jersey, the other in western Pennsylvania. Merton and his associates asked people in these communities to name their three closest friends, and used the resulting data to offer generalizations about the social forces that affect the formation of friendship. Close friendships were most common between those of the same

ethnicity and same gender, a finding that led Merton to propose
a term for a tendency that had been observed millennia earlier, by
Aristotle, in his *Nicomachean Ethics*: "Some define friendship as
a kind of likeness and say like people are friends, whence come
the sayings 'like to like,' 'birds of a feather flock together.'"

Merton coined the term "homophily"—love of the same—to
describe the phenomenon. It wasn't especially surprising to Merton that there were few friendships between black residents and
white residents of these communities, though he expressly chose as
one of his research sites a racially integrated housing project. More
surprising was Merton's discovery of homophily effects around
beliefs and values—people who had similar points of view about
racial cohabitation in housing projects were significantly more
likely to be friends than those who had differing opinions.

Since Merton coined the term, sociologists have seen homophily effects when examining social relations as intimate as marriage and as loose as the exchange of work information between
professional colleagues, or the appearance of people together in
a public place. Researchers document homophily effects around
ethnicity, gender, age, religion, education, occupation, and social
class. The phenomenon so pervades our lives that the authors of
a survey paper, generalizing from dozens of sociological papers,
characterize homophily as "a basic organizing principle" of human
societies and groups.[47]

We seem capable of sorting ourselves—unconsciously, for the
most part—by the most trivial of factors. A recent paper by the
Canadian researcher Sean MacKinnon demonstrates that individuals are more likely to choose seats in a computer lab or a
lecture classroom next to people who resembled their appearance
in terms of hair length and color and whether or not they wear
glasses. Questioned by MacKinnon in a follow-up study, students explained their choices in terms of perceived attitudes; they
believed students who were similar in appearance to them would

be more likely to share their attitudes and more likely to accept and befriend them.[48]

Considering the power of these homophily effects can make us uneasy. The educational psychologist and university president Beverly Tatum titled a book on the development of racial identity *"Why Are All the Black Kids Sitting Together in the Cafeteria?"* to point to the unease we feel when we witness self-segregation. (Tatum's book argues that this sort of self-segregation is necessary for students to become confident in their racial identities, a precursor to building strong friendships across racial lines.) A common personal reaction to the sociological research on homophily is to conduct an internal inventory of our friends to find counterexamples, seeking evidence that we're less subject to homophily effects than the average individual. Most of us think of ourselves as open-minded and unbiased, willing and able to have rewarding social relationships with people of all backgrounds, and seeing contradictory evidence troubles us.

It's a mistake to extrapolate from research on homophily and conclude that "everyone's a little bit racist," as Robert Lopez and Jeff Marx do in a song from their wonderfully transgressive musical *Avenue Q*. If you grow up in a community where there's little racial diversity, your population of possible friends is limited. Sociologists call this "baseline homophily." People develop friendships with people they interact with in a shared activity. If you're an ice hockey player, you're going to meet a lot of white guys from cold climates. Play cricket, and your pool of possible friends will likely look very different. When we examine online activity, we see similar results.

The sociologists Andreas Wimmer and Kevin Lewis used a huge set of data from Facebook—the complete set of postings for one class at one major university for a year—to study racial homophily online and in the physical world. They focused on photos that the students had posted. Students who appear together in a Facebook photo are more likely to be "real" friends than people who've merely "friended" each other on Facebook, a behavior so common at most universities that it's not a strong signal of actual friendship. Wimmer and Lewis saw significant homophily effects and were able to study homophily at a much finer level than in previous studies. They conclude that certain types of homophily—for example, the tendency of Asians to befriend one another—are simply generalized effects from more specific homophilies: the tendency of Indian or Chinese students to befriend one another. And they see distinctive nonethnic homophilies, including a strong tendency for students from Illinois, or math majors, or boarding school classmates,[49] to be friends with one another.

This finding suggests that the effects of homophily come at least as much from structural factors—Whom did you have the chance to meet in high school? In the computer lab?—than from personal choice. The most powerful effect Wimmer and Lewis found in their model was a "closure" effect, a phenomenon described by the sociologist Georg Simmel early in the past century.[50] If Jim is friends with Bob and with Sue, Bob and Sue are likely to become friends, closing the circle of friendship. If Jim is African, there's a higher likelihood than random that Bob and Sue are African. Their friendship isn't the result of each seeking out the comfortable friendship of a fellow African as much as it's a product of social closure. Closure leads to the amplification of other homophily effects, and Wimmer and Lewis's model suggests that it can quickly lead to all the black kids' sitting at the same table. (They would point out that, if we look closely, we'll likely see Nigerians sitting with Nigerians, African Americans from Atlanta with fellow southerners.)[51]

In other words, if you discover that your social circle is highly homogenous in terms of ethnicity, gender, or national origin, it doesn't necessarily mean you're racist, sexist, or nationalist. It more likely means that your circle of friends has been shaped by where you've lived, where you've gone to school, and what interests you've pursued. And while this may not be a surprise to anyone whose Facebook feed looks less like the United Nations and more like a family reunion, it poses a challenge to universities that see part of their educational role as preparing students for a multicultural world.

That there are deep structural factors that help explain homophily doesn't mean it's inevitable, however. The single biggest factor that predicted friendship in Wimmer and Lewis's set was students sharing a dorm room. The university they studied appears to have a policy designed to increase racial mixing. White students were rarely placed in rooms with other white students, for instance. Wimmer and Lewis reach an unambiguous conclusion about this policy: if it's intended to increase the number of friendships that cross racial barriers, it succeeds.

Homophily offers a reminder that our view of the world is local, incomplete, and inevitably biased. Our knowledge of other parts of the world, and our interest in stories from other nations, is influenced by the people we know and care about, and those people are more likely to be our countrymen than people from a different continent.

Just as embracing the assumption that Europe is inexorably sliding to life under sharia law makes it difficult to have a productive conversation about immigration reform, imagining that we have a broader picture of the world than we actually do subjects us to unhelpful distortions and misperceptions. We may feel ready to do business with our new Japanese customers, but there's a good chance that we're simply imagining ourselves as cosmopolitan. We need to ask whether we're reading the *Times of India*, or imagining we are simply because we could be. We have to look less at what's made possible by the Internet and more at what we actually choose to do.

Wimmer and Lewis's study of homophily is ultimately a hope-

ful one. When universities decide their students would benefit from building friendships across racial lines, they've found a
structural solution that works: forcing very different people to
share the same room. If we want to change the information we're
getting about the wider world—to move from imaginary to digital cosmopolitanism—we can make structural changes, too. But
first, we need to take a close look at how we encounter the world
through media.

Chapter 3

WHEN WHAT WE KNOW
IS WHOM WE KNOW

BEFORE 2000, WE ENCOUNTERED NEWS PRIMARILY THROUGH professional curators. For the decade that followed, we began acting as our own filters, searching for what we wanted to know. This decade offers the promise that our friends will help us find what we need to know.

Each of these shifts in media has subtly changed our picture of the world. As we've moved from edited, curated media into search and social media, we have increasing choice over the picture of the world we assemble for ourselves. We also have increased responsibility for building a picture of the world that's accurate and comprehensive enough that we can navigate threats and seize opportunities.

These shifts are not complete or exclusive; we've always built a picture of the world through some combination of our encounters with media, our individual search for knowledge, and our encounters with others. But a shift in volume, from a world in which much of our knowledge and interest comes through curated and edited media, to one where much more comes through search and social, offers a good incentive to consider the strengths and weaknesses of each of these three ways of seeing the world.

I began to think about these different ways of understanding the world in 2000, when I began commuting between continents. I lived, as I still do, in rural western Massachusetts, and I worked in the Osu neighborhood of Accra, the capital of Ghana. When all went well, it took about twenty-four hours to get to work, passing through Boston and Amsterdam en route. All seldom went well.

I was working with friends to start a technology volunteer corps, called Geekcorps, which matched experienced computer programmers with Ghanaian software firms. While computer science is a well-established academic discipline taught in universities throughout the world, software engineering is a craft learned through apprenticeship, working on projects alongside more experienced programmers. Ghana, where I'd lived as a student in 1993 and 1994, had no shortage of smart entrepreneurs who wanted to start web businesses. It did have a shortage of experienced programmers to learn from, so we recruited programmers and graphic designers from the Americas and Europe who were willing to share their talents for months at a time in exchange for food, housing, and the chance to live in Accra, an impossibly colorful city with warm beaches, spicy food, and the atmosphere of a street party virtually every weekend.

Working on Geekcorps required and encouraged a particular strain of optimism. Ghana, blessed with natural and cultural riches, has been slow to develop economically. Economists who study the reasons nations become rich or poor are fond of pointing out that in 1957 (the year Ghana gained independence from the United Kingdom), South Korea and Ghana had roughly the same per capita income. Thirty years later, South Korea was a middle-income nation on its way to becoming a manufacturing powerhouse, whereas incomes in Ghana had actually fallen. Like most African countries, Ghana was having a hard time transitioning from an agricultural economy to an industrial one. Embracing the idea that Ghanaian software firms could jump-start the country's service sector by helping Ghanaian businesses sell to international audiences or take on outsourced tasks from companies in the United States and Europe required something of a leap of faith.

Yet there were good reasons to believe this scenario was possible. Ghana's national language is English, and many Ghanaians are well educated, a strong combination for winning outsourcing jobs. When we began working in Accra, Ghanaians were working for Data Management International (DMI), processing New York City's parking tickets, translating the scrawls of police officers into neat, searchable databases.[1] With plans underway for a massive underwater cable connecting Ghana to the Internet via fiber, rather than via satellite, it seemed possible that Ghana might become a hub for English-language call centers and challenge India's dominance in the data entry space.

But projects like DMI's were few and far between. Potential customers had just gotten comfortable with outsourcing business to India—Ghana was a step too far for most. More common than interest in Ghana was skepticism on the part of foreign investors and reluctance on the part of government and international aid contractors to work with local firms. For investors, it was safer to put money in markets closer to home. For aid agencies, it was safer to contract with US and European companies that had long track records rather than with local firms with less experience. For Ghana to attract investment and new business, it wasn't enough to be optimistic about Ghana's future; that optimism needed to be exported and spread.

My commute between continents gave me ample time to read, and I boarded monthly flights armed with a stack of newspapers and magazines. I'd read the *Economist*, the *Guardian*, and the *New York Times* on my way there, leave my copies with Ghanaian friends, and return armed with the *Daily Graphic*, the *Accra Mail*, *African Business*, and the *New African*. The African papers weren't as well written or professionally produced as the New York and London papers, but they covered local, regional, and international news. The American and UK papers, for the most part, appeared to have a blind spot that blotted out most of the African continent.

On December 28, 2000, John Kufuor won a runoff election and became president-elect of Ghana. The result was unprecedented in Ghanaian history, and extremely rare at that point in African history: an opposition candidate won a free, fair, and largely peaceful election, defeating a candidate backed by the former dictator, who had voluntarily relinquished his power. As someone who had become professionally interested in good news from Africa, I expected the story of Kufuor's election and Ghana's dramatic change to be front-page news in the United States.

At home for the holidays, I flipped through the *New York Times*, looking for American recognition of this rare, optimistic African story. When I finally found a 300-word story on the election buried deep in the paper, I was infuriated. No wonder we couldn't get American companies to take the idea of business in Ghana seriously. They weren't hearing any of the good news coming from the continent. In fact, they weren't hearing about Africa at all.

I can now recognize that the *Times* did a pretty good job of covering Kufuor's election. It ran four brief stories on the elections over a week and, ten days after the election, a laudable editorial, titled "An African Success Story," which praised Ghana in just the way I'd hoped.[2] At the time, however, the apparently scarce coverage sent me searching for the wisdom of hard-bitten journalists frustrated by the shortcomings of international media. I found an article by the journalist Peter Boyer, whose angry reflection on media coverage of the 1984–85 Ethiopian famine included this observation: "One dead fireman in Brooklyn is worth five English bobbies, who are worth 50 Arabs, who are worth 500 Africans."[3]

I decided to see for myself whether Boyer's observation was true, that newspapers were systematically underreporting news from Africa. As it turns out, I would have been better off searching the academic literature. George Washington University professor William Adams tested Boyer's formulation in a 1986 paper called "Whose Lives Count? TV Coverage of Natural Disasters."[4] He concludes that US television coverage of natural disasters correlates to US tourist traffic to a country, a nation's proximity to the United States, and the size of a disaster, but cultural-proximity factors outweigh the magnitude of the tragedy: the equivalent earthquake in Canada and Cameroon will likely receive very different levels of attention, as American audiences are more culturally connected to Canada than to Cameroon. But during my days with Geekcorps I wasn't even pretending to be an academic, so I started writing code, assuming that none of the world's media scholars had noticed the gap in attention between Africa and the rest of the world.

Mapping Media Attention

My experiment was less sophisticated than Adams's, but broader in scope. In 2003, I wrote a series of software scripts that polled the websites of major media outlets—the *New York Times*, the BBC, Google News, and a dozen others—each day and searched for stories that mentioned the country or the capital city of over two hundred countries and territories. Using this data, I produced daily maps that offered a summary of what nations were receiving more or less media coverage. And, like Adams, I used statistical techniques to figure out what factors best explain which countries are in the news and which get short shrift.

The maps I produced were quite similar across media and across time. An attention map of the *New York Times* looks a great deal like one of Google News, though one represents a single source while the other aggregates thousands of sources. Similarly, it's difficult to tell a map from 2003 from one made in 2007 as the same general patterns hold. American media pay

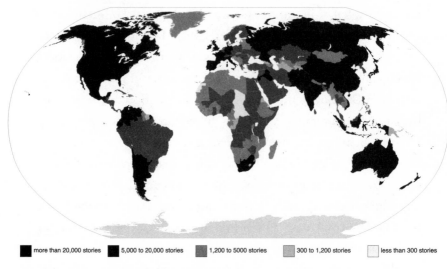

more than 20,000 stories   5,000 to 20,000 stories   1,200 to 5000 stories   300 to 1,200 stories   less than 300 stories

Google News Media Attention, February 2004.

close attention to western Europe, to the large Asian economies (China, Japan, and India), and to major powers in the Middle East. Sub-Saharan Africa, eastern Europe, Central Asia, and South America all receive much less attention.

Mapping media in this fashion raises an awkward question: What's the appropriate level of media coverage for a nation to receive? Boyer's critique suggests that we might aim for an ideal world where each person's life or death is equally newsworthy, be they English, Arab, or African. But that idea doesn't scale very well. There are 4,116 people in China for every one in Iceland. We would need several years of news about Chinese citizens before we received a single dispatch from Reykjavik. Should news coverage be proportional to negative impacts, the number of individual deaths? To surprise? Should coverage favor news that might impact a reader or viewer—events in countries with strong economic or cultural ties to our own?

Deciding what's news is both a thorny philosophical question and an intensely practical one: it's what editors do in newsrooms every day. Given the difficulty in defining what factors should contribute to newsworthiness, it's worth looking at what factors—explicit or implicit—help explain what and who make the news.

Disparities in media attention can be massive, even between countries with similar populations. Japan's population is just over 127 million and is slowly shrinking. Nigeria's population surpassed Japan's in 2002, and now exceeds 154 million, ranking it seventh among the most populous nations. There's no shortage of newsworthy events in either country, but their media footprint is vastly different. In an average month, US publications will run roughly eight to twelve times as many stories that mention Japan as stories that mention Nigeria.

If attention were proportional to population, we would see a rising interest in Nigeria as it grows and Japan shrinks. If common language, religion, or geographic proximity were the key factor, we would still expect to see more Nigerian news, since Nigeria and the United States have a common language and a large Christian population and are physically closer than the United States and Japan. If conflict drove coverage, we would, again, expect to find more Nigerian stories. The country has suffered ethnic and religious violence and terror attacks in the past few years, while Japan has been free of civil strife (though profoundly affected by natural disasters in 2011, well after I'd completed this study).

The single factor that best explained the distribution of media attention in US sources that I tracked from 2003 to 2007 was gross domestic product. For some US news sources, GDP explained 60 percent of the variation in news coverage. Japan is the world's third-largest economy, while Nigeria ranks forty-first, behind much smaller nations like Finland and Denmark. While I find it reassuring that there's an explanation for Nigeria's comparative invisibility that's more quantifiable than racism or Afrophobia, it's disconcerting to conclude that US media have such a pronounced financial fetish and so little interest in poorer nations.[5]

The other major determining factor for US media attention is the involvement of the US military. Iraq and Afghanistan, both poor, were heavily covered in US media during the years of my study. And countries that rarely register in the US media can vault to prominence when the military gets involved. When US Marines landed in Monrovia, Liberia, to help end the second Liberian civil war, in August 2003, the nation previously invisible on my maps was at the peak level of media attention for about two weeks, then receded.

The "GDP + US troops" model isn't universal. For example, the BBC's attention profile was better explained by a model that considered GDP and Britain's colonial legacy. Nations like Kenya, Zimbabwe, South Africa, and India are more prominent on the BBC than on American media outlets, while parts of the world where Britain had far less colonial influence, like South America, are less prominent. And, just as William Adams also saw in his study, countries show a strong interest in their neighbors. Rivals are also well represented: the *Times of India* features heavy coverage of Pakistan, India's chief military rival, and China, its chief economic rival. Whether a media agenda oriented toward a colonial past or focused on current rivals is any more fair than an agenda tightly correlated to national wealth is not clear, but the underlying biases and assumptions are.

Late in 2003, I published a paper on my findings and started posting my maps to my website. Within a few weeks, I received an email from a journalism professor who politely wondered whether I realized that there was a long history of research and activism around media attention. She suggested I start with a paper from 1965 by Johan Galtung and his graduate student Mari Ruge, "The Structure of Foreign News."

Galtung is a Norwegian sociologist whose six-decade-long academic career has focused on questions of peace and conflict. He founded the *Journal of Peace Research* in 1964 and the Peace Research Institute Oslo, two of the central institutions in the peace and conflict studies community. Early in his research, Galtung scrutinized the effect of media on peace and conflict. He and I therefore shared a pastime: the obsessive counting of foreign news stories in newspapers.

"The Structure of Foreign News" examines four years of coverage (1960–64) by four Norwegian newspapers, looking at stories about three international crises, in Congo, Cuba, and Cyprus. Galtung's goal in analyzing these stories was to determine "how do 'events' become 'news'?" He and Ruge suggest that we "tune in" on events in the same way that radio receivers tune in to signals amid the noisy mishmash of radio waves. If we're turning the dials (remember, this is an analog metaphor) on a shortwave radio, we're more likely to pay attention to signals that are clear, loud, and meaningful and miss those that are noisy and unfamiliar. By analogy, Galtung and Ruge offer a set of "news values" that describe events we're likely to register as news.

News, they suggest, has a frequency: events that transpire over a very long time, like climate change or economic growth, are less likely to become news than those that happen within a twenty-four-hour news cycle—a tornado or a stock market crash. Events that are unambiguous—good or bad—are more likely to become news, as are ones that are more meaningful to us, in terms of cultural proximity, relevance, or comprehensibility. They suggest that newsworthy events swing between unexpectedness and consonance. Unexpected events are more likely to be newsworthy than commonplace ones: man bites dog is news, while dog bites man is not. At the same time, news is likely to reflect our preconceptions. We're more likely to get news of conflict or famine in Africa than an unexpected story about business opportunity. Galtung and Ruge see evidence to support Boyer's hypothesis that the media seem to deem some lives more worthy of attention than others. They see "topdog" nations receiving more coverage than "underdog" nations, and elite persons, like leaders and celebrities, receiving more attention than average citizens.

The dozen news values that Galtung and Ruge offer can help

us analyze current news imbalances. We might explain a systematic bias in attention to Japan over Nigeria in terms of reference to elite nations, or cultural proximity, while the light coverage of Ghana's presidential election might suggest that the story fails the test of predictability or reference to something negative, by being unexpectedly positive. The paper is more helpful in proposing these factors than in proving their influence on Norwegian news coverage in the early 1960s. And Galtung and Ruge aren't clear on whether these factors are actively on the minds of news editors or are instead unconscious biases.

They are convinced, however, that the media of their day are obstacles to peace: "Conflict will be emphasized, conciliation not." And they worry that an emphasis on powerful nations—at the time, the United States and the Soviet Union—suggests a frame for news where events are analyzed in terms of whether they're good or bad for the West, not whether they're good or bad for the people directly affected by them.

Agenda Setting

Why would it matter that we hear more about Japan than about Nigeria, more about American military intervention than about progress made by African democracies? In 1963, the political scientist Bernard C. Cohen offered one answer: the press "may not be successful much of the time in telling people what to think, but it is stunningly successful in telling its readers what to think about." The journalism scholars Maxwell McCombs and Donald Shaw called Cohen's idea "agenda-setting," and set out to test it, surveying voters in the 1968 US presidential election and closely analyzing the newspapers and television programs they would be most likely to access. McCombs and Shaw found a strong correlation between issues voters identified as the most important in the 1968 campaign and issues that received heavy coverage in local and national media. It's possible that newspapers were reacting to the interests of the readers, but that scenario is hard to imagine, given the limited ability for newspapers to track which stories in an issue were read and which were ignored. It seems more likely that the news outlets in 1968 were promoting the importance of some issues over others and that those choices shaped what issues voters considered to be important.[6]

Agenda setting, like many important ideas, seems obvious in retrospect. News signals what's important. Because a near infinity of events occurs every day, we would be overwhelmed by a stream of non-newsworthy events: every city council meeting, every parliamentary debate, every petty crime. We need someone or something to tell us what events should be considered news, and whoever makes that choice has tremendous power. It's hard to get outraged over the local government decision you don't hear about, or to mount a campaign to right an injustice you've not learned of. Whoever chooses what's news has the power to influence our cognitive agendas, to shape what we think about and don't think about.

In a book examining US media coverage of the Vietnam War, the political scientist Daniel Hallin offered a deceptively simple diagram to explain some of the implications of agenda setting. The diagram, sometimes called "Hallin's spheres," offers a circle within a circle, floating in space. The inner circle is the "sphere of consensus," which Hallin explains is "the region of 'motherhood and apple pie'; it encompasses those social objects not regarded by journalists or most of society as controversial." It's surrounded by a larger "sphere of legitimate debate," issues on which it's well known that "reasonable people" may have different views.

In the United States, the idea that representative democracy and capitalism are the correct organizing principles for modern society is within the sphere of consensus. You'll see little journalistic coverage of arguments that the United States should engage in a socialist redistribution of wealth or become part of a global Islamic caliphate. The sphere of legitimate debate includes conflicts over abortion rights, restrictions on firearm ownership, or levels of taxation. Stray away from issues where there's consensus, or consensus that there's a debate, and you find yourself in the "sphere of deviance," where points of view aren't even considered part of the media dialogue. Hallin observes that the "fairness doctrine," a policy of the US Federal Communications Commission (FCC) from 1949 to 1987 that required broadcasters to devote substantial time to public issues and to ensure representation of opposing views, explicitly stated that broadcasters were not required to give airtime to communists.

The press, Hallin argues, plays a role in "exposing, condemning, or excluding from the public agenda" the deviant views. By reporting some views and not others, the press "marks out and defends the limits of acceptable political conduct." Even if, among scientists, human impact on the climate has moved into sphere of consensus, as long as American media keep airing dueling points of view about the issue, it will remain within the sphere of legitimate debate. By covering "birther" accusations that President Obama's birth certificate was invalid, the press moved a previously deviant idea into the sphere of legitimate controversy and turned a conspiracy theory into a major political debate.

Viewpoints don't need to be particularly distasteful or offensive to enter the sphere of deviance; they simply need to be far enough outside the mainstream that "serious people" won't engage them. The political cartoonist Ted Rall proposes a simple test for detecting deviance: "When 'serious people say' something, those who disagree are by definition trivial, insipid and thus unworthy of consideration. 'No one seriously thinks' is brutarian to the point of Orwellian: anyone who expresses the thought in question literally does not exist. He or she is an Unperson."[7] The phrase "serious people" is also an attempt to defend the journal-

ists' notion of the sphere of legitimate controversy against ideas
from the sphere of deviance.

The NYU journalism professor Jay Rosen argues that Hallin's
spheres help explain public dissatisfaction with journalism. "Anyone whose basic views lie outside the sphere of consensus will
experience the press not just as biased but savagely so." Believe
that separation of church and state is a poor idea or that the government should be the primary provider of health care services,
and you'll be so far outside the US sphere of legitimate debate
that you'll never see your opinions taken seriously in the mainstream press, which may leave you alienated, disaffected, and
looking for other sources of news.[8]

The sphere of deviance is also the sphere of obscurity. Believing
that Nigeria is as newsworthy as Japan is probably as far outside
the sphere of legitimate debate in the United States as advocating
for single-payer health care. Rosen points out that these spheres
are political, not in the sense of left/right, Republican/Democrat,
but in the sense of defining what's worth the public's time and
attention. The struggle to get "serious people" talking about an
issue—be it responses to famine in the Horn of Africa, or doubt
that the US president is an American citizen—is the political
struggle to take an issue from obscurity and turn it into a valid
topic of debate.

The Gatekeepers

When Boyer complains that not all human lives are reported
equally, when Galtung proposes news values, Hallin spheres of
coverage, or McCombs and Shaw agenda setting, they're all placing the responsibility and blame on editors and publishers. Editors and publishers are the "gatekeepers" who decide what stories
receive coverage and, indirectly, what ideas are the object of public debate.

The term "gatekeeper" was coined by the Prussian social theorist Kurt Lewin in 1947. He wasn't writing about making newspapers; he was trying to get American housewives to change what they served for dinner. Lewin's research on the topic was sponsored by the US government, which wanted to encourage increased consumption of "secondary cuts" of beef—organ meats, tripe, sweetbreads—so that primary cuts could go to feed soldiers. Would lectures on the virtues of beef hearts, delivered to Iowa housewives, change their purchasing behavior? Lewin identified "channels" that bring food to a family dinner table and gatekeepers who controlled inputs into these streams. Housewives, in his analysis, were the ultimate gatekeepers over what families ate for dinner.[9]

Lewin didn't live long enough to extend his theories beyond their narrow remit; he died of a heart attack before the publication of his initial gatekeeping paper. His student David Manning White brought the theory of gatekeeping into the world of journalism in 1949, analyzing the decisions an editor named Mr. Gates made at an Illinois paper, the Peoria *Star*, in choosing what stories offered by reporters and wire services made it into the newspaper. White saw Mr. Gates's process as highly personal and idiosyncratic. Rather than following the lead of larger city papers, Gates choose stories he was personally interested in and thought his readers would find interesting as well. Paul Snider visited Mr. Gates seventeen years later, in 1966, and reported that his selections remained similar. With the rise of the Vietnam War, Mr. Gates featured a bit more international news, but selected "a balanced diet" of events and personalities to meet the tastes of his readers.[10]

Whereas White saw gatekeepers as makers of personal decisions about what constituted news, the journalism scholar Walter Gieber argued that gatekeepers were less like Mr. Gates and more like cogs in a machine. Gieber studied the gatekeeping decisions made by sixteen news service editors and concluded that they were "concerned with goals of production, bureaucratic routine and interpersonal relations within the newsroom." Their judgments were less about personal, subjective applications of news values and more a response to the constraints of the structures that governed their work.[11]

If Gieber is right, and gatekeepers are constrained by the structures they work within, creating media that are more representative or more hopeful about the future of Africa requires not just changing Mr. Gates's idiosyncratic mind but also altering the systems he's embedded in. That was the goal of a commission convened by the United Nations Educational, Scientific, and Cultural Organization (UNESCO) in 1977 to address the challenges of communications in an interconnected world. Amadou-Mahtar M'Bow, the UNESCO director general who convened the commission, sounds like an early digital cosmopolitan in his framing of the problem:

> Every nation now forms part of the day-to-day reality of every other nation. Though it may not have a real awareness of its solidarity, the world continues to become increasingly interdependent. This interdependence, however, goes hand in hand with a host of imbalances and sometimes gives rise to grave inequalities, leading to the misunderstandings and manifold hotbeds of tension which combine to keep the world in ferment.

Scholars from sixteen nations, led by the Irish Nobel Peace Prize recipient Seán MacBride, considered a huge set of these imbalances: the geographic distribution of communication technologies, from printing presses to communications satellites; the flows of television shows and movies from the United States to developing nations; ownership of news wire services by American and European firms and a perceived bias against news from developing nations. The commission produced eighty-two recommendations, ranging from the quotidian (increasing the international paper supply to make newsprint cheaper) to the fanciful (a satellite network to "enable the United Nations to follow more closely world affairs and transmit its message more effectively to all the peoples of the earth").[12] Many of the suggestions explicitly addressed media imbalances: "the media in developed countries—especially the 'gatekeepers,' editors and producers of print and broadcasting media who select the news items to be published or broadcast—should become more familiar with the cultures and conditions in developing countries."[13]

The MacBride commission chose an unfortunately Orwellian name when it released its proposed agenda in 1980: the New World Information and Communication Order (NWICO). The MacBride report was viewed by some readers as endorsing increased state control over news and restricting press freedom. While the text of the report forcefully defended rights of press freedom, with statements like "Censorship or arbitrary control of information should be abolished," the apparent alignment of the USSR with developing world representatives on the commission led American and British commentators to see the Mac-Bride report in a different light. They argued that it proposed the licensing of journalists by governments and the support of state news agencies to compete with private outlets. By 1983, the *New York Times* had condemned the report on its editorial page and endorsed a US withdrawal from UNESCO. The United States and the UK left UNESCO in 1984 and 1985 in protest over NWICO, and neither country rejoined for more than a decade.

The Power of the Audience

Whether the MacBride proposals were a veiled attempt at authoritarian control of the press or well-meaning but impractical suggestions for addressing media imbalance, their rejection highlighted the difficulty of using international mandates to change the behavior of gatekeepers. Armed with my data, historical research, an overdeveloped sense of self-righteousness, and a limited understanding of the business side of journalism, I decided to confront gatekeepers individually over their coverage decisions.

My first target was Jon Meacham, then the managing editor of *Newsweek*. At a meeting of the World Economic Forum in Switzerland in 2004, I took advantage of the question-and-answer period after his talk to launch a jeremiad condemning American media's undercoverage of international news. Meacham responded graciously to my broadside, asking, "You realize we're on the same side, right?" He'd love *Newsweek* to publish more international news, he explained, but every time he put an international story on the cover, newsstand sales dropped sharply. "You and I may both want more international news, but it's not clear that our readers do."

The numbers are with Meacham, at least as regards American audiences. In a recent poll, 63 percent of the respondents said that they were getting enough international news and wanted to hear more local stories. Meacham's comment reminds us that the "Chinese wall" between the business and editorial sides of for-profit news organizations is a flimsy and transparent one. In a predigital age, it was hard to tell which stories were attracting the attention of newspaper or magazine readers, except in the case of cover stories. But in a digital age, there's a wealth of data about which stories are read and which are ignored.

The news industry analyst Ken Doctor explains that while these numbers were traditionally shared with reporters on a "need to know" basis, the *Washington Post* is now sending three traffic reports an hour to 120 staff members, giving them up-to-date information on which stories are rising and falling in popularity.[14] Other news outlets are less shy about delivering metrics to their writers. The *Huffington Post* relentlessly optimizes its front page, on the basis of the traffic that stories are receiving, and reporters are acutely aware of how much attention their stories are getting.[15] An extreme case, Gawker Media, proprietor of several prominent "lifestyle news" websites, decorates its newsroom with the "Big Board," a computer monitor that displays which stories are receiving the highest traffic, updated in real time, to help incent employees to publish stories that reach a wide audience.

When news was delivered on paper, instead of on computer screens, editors relied on their judgment to determine which stories were likely to generate "MEGO." William Safire explains this important industry term: "MEGO is an acronym coined by *Newsweek* staffers for 'my eyes glaze over,' to describe audience reaction to subjects that everyone agrees are important but are surefire soporifics. Latin American policy, Eurodollars, and manpower training are MEGOs. A speech on government reorganization, written and delivered in a monotone, can achieve the 10-point MEGO on the Richter scale, putting the entire audience to sleep."[16] Today's enhanced metrics deliver real-time evidence of MEGO and immediately penalize attempts to write about serious international issues that fail to engage audiences.

Given this new real-time feedback environment, the mystery may not be that we get so little international news but that we still get so much. Sending reporters oversees to cover events is expensive, and commercial news outlets have been shrinking their foreign presence. The American network CBS had thirty-eight correspondents in twenty-eight bureaus in the mid-1980s and had reduced its footprint to five bureaus in four countries by 2008.[17] Major US newspapers like the *Boston Globe* and *Baltimore Sun* eliminated their overseas bureaus, which had been a source of pride and prestige for the papers, as cost-cutting measures.[18] Much of the international news that appears in smaller American newspapers comes from wire services, which means the stories will appear in hundreds of other papers. Correspondents on the ground allowed newspapers like the *Globe* to offer exclusive international content, whereas running a Reuters story on devel-

opments in Afghanistan provides no differentiation or competitive advantage for a city newspaper.

Four major American newspapers still maintain substantial foreign bureaus: the *New York Times*, the *Los Angeles Times*, the
Washington Post, and the *Wall Street Journal*. The *New York Times*
and the *Journal* reserve roughly 20 percent of their total news space
for purely international stories; each allocates another 10 percent
to stories on US foreign affairs.[19] It's possible that these papers publish so much more international news than smaller US papers in an
effort to position themselves as "elite" media outlets, required reading for those engaged with a world larger than their hometown. Or
perhaps the editors of these papers are neo–Mr. Gateses, fighting
market pressures more successfully than Meacham had been, ignoring evidence of what the audience demands and delivering what
they believe to be most newsworthy and civically relevant. Without
access to their traffic statistics, it's hard to know.[20]

The use of analytics by gatekeepers grants a new form of
power to the audience. When readers signal their interest in an
international story, they often are rewarded with more coverage.
Heavy focus on the Darfur conflict in US newspapers, and on
Zimbabwean elections in UK papers, suggests that engaged con-
stituencies can keep stories in the newspaper through interest and
feedback, perhaps at the expense of stories with more import but
less audience, like the conflict in eastern Congo. We can blame
the gatekeepers, but we must also examine our role, individually
and collectively, as audience.

Be Your Own Gatekeeper: The Arrival of "The Daily Me"

The power of the digital age makes it as easy to access one news
source as another, and to pick and choose the news we want to
see. This new arena, with its low cost barriers to publishing and

essentially unlimited space, means news isn't constrained by the bundle of paper delivered to our doorstop or purchased at a corner shop, nor does it depend on transmission over a limited, crowded broadcast spectrum. Rather than letting professional gatekeepers, hobbled by business concerns and dominated by the biases of their news values, govern what's in our sphere of legitimate debate, we can seek out the news we want and need.

We no longer have to rely on newspaper editors to curate a front page of news we "need to know"—we are our own curators, selecting what we want and need from a near-infinite range of possibilities. This new power makes the idea that international news competes with domestic reporting, sports, celebrity gossip, and advertising for precious column inches in a newspaper, or time during a television broadcast, seem decidedly twentieth century. But how did we get here?

Somewhere in the mid-1990s, a dramatic shift occurred in the way we organize information, a shift from curation to search. Curators—to use a new term that came to encompass all manner of professional gatekeepers, from editors to news anchors to media critics—grew less powerful and were subject to greater critique and scrutiny, while new, powerful organizations were built around the power of search.

People quickly grew accustomed to the idea that they could use a search engine to discover information on any topic of interest. Exploring the Internet moved from directionless "surfing" to goal-oriented searching. Being able to find exactly what you wanted to know invites you to question authority figures—editors, educators, doctors—who argue there are topics you need to know beyond those you want to explore. Companies like Google realized that a conceptual shift was underway and built a business around the idea that you knew what you wanted to know better than any expert ever could.

I had a front-row seat for this shift, which I watched, baffled and disconcerted, as I helped build a company in the early days of the commercial Internet. From 1994 to 1999, I was the head of research and development for a website called Tripod.com. When I joined the company, Tripod's mission was to provide high-quality, edited content for recent college graduates, helping them land jobs, rent apartments, fall in love, and generally achieve twenty-something happiness. We wrote stories, published guides to the best content online, and applied our curatorial intelligence to the rapidly expanding world of digital media.

Late one night in 1995, one of my programmers—Jeff Vander Clute—had a clever idea. Earlier, we'd built a simple tool that allowed people to enter information into a form and produce a formatted résumé, essentially a web page that lived on our server. Jeff realized we could put up a much simpler form—a big blank box—and allow people to upload whatever web pages they chose to build. He wrote up the code, we called it "The Homepage Builder," put it on the server, and promptly forgot about it.

I didn't think about the homepage builder for nine months, until I got a call from our Internet service provider, who informed me that his bill for hosting our site had increased by a factor of ten. I demanded an explanation, and he responded with charts of our bandwidth usage. We'd gone from hosting a few thousand visitors a day to hosting hundreds of thousands. I hadn't noticed. I had been diligently monitoring traffic to the content we'd painstakingly written and put online, and not monitoring whatever it was our users were creating. Those pages now represented the vast majority of our site, in terms of total pages and traffic.

Our business model had been based on paying professional editors to create web pages and selling expensive ads on those pages. It took roughly eighteen months to figure out that we were in the wrong business. We'd been surprised by two trends: the rise of participation and the rise of search. With tools like our homepage builder, millions of ordinary individuals were joining the tens of thousands of companies that were making content available to readers for free.

Until 1998, it had been unclear whether users would navigate the web by using curated directories, like Yahoo!, or search engines, like Altavista or Lycos. Google's ascendancy, which began in 1999, paralleled an explosion in content created by the millions of people publishing online. The quality of the average page on the web plummeted, but the total amount of worthwhile information increased sharply. It just was much harder to find. As the size of the Internet exploded from several million web pages into many billion, search rapidly emerged as the only practical way to navigate this ocean of content.

Tripod's readers weren't interested in the articles we'd carefully crafted for them. They were coming to explore thousands of topics we knew nothing about: Malaysian politics, Japanese animation, customized cars. We thought we were running a newspaper for recent college graduates, telling them what they needed to know to succeed in the world. They helpfully told us that they couldn't care less—if there were topics they wanted to know about, they'd find them through search, and they cared very little whether those stories were written by professional authors or ordinary Internet users.

Once you've discovered that what interests you might be found in any corner of the Internet, a general-interest news source makes little sense.

If the promise of a high-quality newspaper is that you'll find everything you *need* to know about the day's news within its pages, the promise of search is more seductive: somewhere on the Internet is everything you *want* to know, and we can help you find it, with a minimum of what you don't want.

One of the thinkers who first recognized the implications of the rise of search and the fall of curation for newspapers was Pascal Chesnais. Working with a team of researchers at MIT in 1994, he introduced a news service called "The Freshman Fishwrap." Drawing from a pool of four thousand stories a day delivered via the Associated Press, Knight-Ridder, and Reuters wire services, the Fishwrap offered a user-customized newspaper with stories about a student's hometown, favorite sports teams, and topics of interest. Rather than the professional editorial judgment of a Mr. Gates, a reader was her own gatekeeper, asking Fishwrap to deliver the stories she wanted to encounter, and to suppress the rest.[21]

Nicholas Negroponte's *Being Digital* doesn't mention Fishwrap by name, but it describes a similar-sounding technology: "What if a newspaper company were willing to put its entire staff at your beck and call for one edition? It would mix headline news with 'less important' stories relating to acquaintances, people you will see tomorrow and places you are about to go to or have just come from. . . . You would consume every bit (so to speak). Call it The Daily Me."

A paragraph later, Negroponte describes a less personalized, more serendipitous newspaper designed for casual reading on Sunday afternoon by the general public, "the Daily Us." But it was the Daily Me that caught public attention, both as inspiration for personal newspapers like MyYahoo! and as the focus of a philosophical critique by the constitutional scholar Cass Sunstein. Sunstein's 2001 book *Republic.com* opens with a chapter titled "The Daily Me," which starts with a speculation about a future world: "The market for news, entertainment, and information has finally been perfected. Consumers are able to see exactly what they want. When the power to filter is unlimited, people can decide, in advance and with perfect accuracy, what they will and will not encounter. They can design something very much like a communications universe of their own choosing."

Sunstein considers this hyper-personalized world a dangerous one, where people's opinions become more extreme in an "echo chamber" of consonant voices. In a subsequent book, *Infotopia*, Sunstein describes an experiment he and colleagues conducted in 2005 to study the phenomenon of group polarization. They invited a set of Colorado citizens from two communities—liberal Boulder and conservative Colorado Springs—to come to local universities and discuss three divisive political topics: global warming, affirmative action, and civil rights. The groups—five to seven citizens selected at random from the same community—had a strong tendency to become more politically polarized over the course of the brief discussion. Liberals became more liberal and conservatives more conservative, and the range of ideological diversity in each group decreased.

Explaining the findings, Sunstein offers multiple possible explanations. In a group setting, people will often gravitate toward a strongly stated opinion, especially if their own opinions aren't fully formed. An ideologically coherent group is likely to repeat a great deal of evidence for one side of an issue and give more reinforcement for that viewpoint, a phenomenon called "confirmation bias." People also find it difficult to defy the will of a group, and some may polarize their views to avoid interpersonal conflict.

Some of these cognitive biases may apply to information encountered online instead of in face-to-face deliberation. Read only right-wing newspapers and blogs, and you'll encounter many strongly stated opinions that may help cement your own. You're likely to encounter lots of information that supports your point of view (confirmation bias) and may encounter few contradictory facts, suggesting that the evidence supports your case (the availability heuristic, where the ready ability to recall evidence that supports your case can blind you to other views). The Daily Me then becomes a machine for polarization, and Sunstein believes he sees the beginning of the Daily Me in blogs: "The rise of blogs makes it all the easier for people to live in echo chambers of their own design. Indeed, some bloggers, and many readers of blogs, live in information cocoons."[22]

Sunstein's writings on polarization are sufficiently controversial that they've generated almost enough academic literature to define a subdiscipline: echo chamber studies. Most responses to his argument don't attempt to counter his theories of polarized deliberation. Instead, they offer evidence that the web's diversity of perspectives prevents people from being overly isolated, even if they're consciously or unconsciously seeking isolation.

The political science professor Henry Farrell and colleagues at George Washington University examined the habits of US blog readers using data from the Cooperative Congressional Election Study, a large social survey conducted by a consortium of thirty-nine universities, and found that blog readers were both unlikely to read blogs across ideological lines and showed much higher political polarization than the average voter.[23] Other studies looked at patterns of links between blogs and found few links across ideologies in the US blogosphere.[24] One study suggests that such links as do exist are often contemptuous, pointing to a differing perspective in order to denounce it.[25]

But readers of political blogs aren't representative of all Internet users. John Horrigan, associate director of research at the Pew Internet and American Life Project, surveyed Americans to see what political arguments they'd heard in the run-up to the 2004 presidential elections. Internet users, they concluded, are more widely exposed to arguments they disagree with than non-Internet users at a similar level of education.[26] And the economists Matthew Gentzkow and Jesse Shapiro used information from an online advertising company to conclude that, while some corners of the Internet may be highly polarized, the websites viewed by the largest audiences draw users from the left and from the right. This paper was warmly received by the newspaper columnist David Brooks, who declared, "If this study is correct, the Internet will not produce a cocooned public square, but a free-wheeling multilayered Mad Max public square."

Gentzkow and Shapiro compared data about users' political preferences from several thousand Internet users with data on those users' visits to 119 large news and politics websites. From this comparison the researchers were able to estimate that, for instance, 98 percent of the viewers of (the conservative commentator's website) rushlimbaugh.com identify as conservative, while only 19 percent of the viewers of (the liberal activist site) moveon.org do. The economists use the "isolation index," long used by sociologists to measure the likelihood of meeting someone from another social group or belief system, to examine the gap between liberals and conservatives visiting a site. The isolation index for a site visited equally by liberals and conservatives would be a zero, while for rushlimbaugh.com, it's 96 (98 percent conservatives minus 2 percent lost and disoriented liberals).

While the isolation index of highly politicized sites seems to support Sunstein's contention of polarized online spaces, the authors also discovered that readers spend a lot of time on sites that have a less polarized audience—general news sites like *Yahoo! News* or *CNN.com*. When you look at the span of sites the average conservative user views, the audience of those sites is 60.6 percent conservative—similar to the audience of (the centrist newspaper website) usatoday.com. Across the span of sites the average liberal visits, the audience for those sites is 53.1 percent conservative. Calculated as the difference between sites popular with conservatives and sites popular with liberals, the isolation index—the difference between those figures—is 7.5, a figure the authors characterize as "small in absolute terms."

Gentzkow and Shapiro go on to compare the Internet isolation index with other types of media in the United States, and discover that local newspapers, national magazines, broadcast television, and cable television all show lower isolation indexes than online media. The only media that show a higher isolation index are "national newspapers," a set that includes only *USA Today* (a centrist, low-prestige paper), the *New York Times* (a left-leaning, elite paper), and the *Wall Street Journal* (a Rupert Murdoch–owned, right-leaning, elite paper). Telling Americans that the readerships of the *New York Times* and the *Wall Street Journal* are politically polarized surprises absolutely no one. Discovering that Internet sites are more politically polarized than cable television news is a surprise, and one that suggests Sunstein's worries might be justified. The economists recalculated their figures by removing the two largest, general-interest sites—*AOL News* and *Yahoo! News*—and found that the isolation index for Internet sites was now higher than for all other off-line media, more polarized than readerships of the *New York Times* and the *Wall Street Journal*.

So why does this research leave David Brooks optimistic that the Internet isn't leading to ideological isolation?

When Gentzkow and Shapiro talk about online isolation as being "small in absolute terms," they are comparing the isolation indexes they're calculating to a measure of homophily in off-line life. Using the 2006 General Social Survey, another massive, multi-university sociological study, they calculate isolation indexes for off-line interactions. Most people report that their close friends and family share their political views. This means we're ideologically isolated when we spend time with friends and family. The economists calculate the isolation index for "trusted friends" as 30.3—that is, if you identify as conservative, you're likely to report that 65 percent of your friends are conservative and 35 percent liberal. That isolation index is more than three times as high as the isolation they see online, and they calculate high isolation indexes for friends, family, co-workers, and neighbors.

In other words, Gentzkow and Shapiro argue you're more likely to encounter someone with a different ideology because you're both reading CNN online than you are to find a neighbor with different political views.

Not so fast. The General Social Survey asks respondents their *perception* of the ideologies of their neighbors. We generally perceive our neighbors to view the world the same way we do—the isolation index for neighborhoods is 18.7. Calculate isolation indexes using actual data on geography and political preference, and neighborhoods turn out to be less homogenous than we think. Calculated this way, the isolation index is 9.4 for zipcode, 5.9 for counties. In other words, I'm more likely to encounter someone of a different ideology in my rural county than I am reading news and opinion websites, because where I live is more ideologically diverse than I perceive it to be.

By comparing their online data with perceptions of homophily off-line, Gentzkow and Shapiro can conclude that Sunstein is overstating his case and that we've yet to fall into a world as segregated and isolated as that of the Daily Me. But their study, like most of the other research studies, suggests that some population of Internet readers are selecting information that's highly partisan politically. Those are probably people highly engaged with politics and frequent readers of political blogs. Still, in their online travels even they are likely to stumble into less partisan spaces, if only to check the sports scores. Larger audiences are finding news in spaces that are less partisan than the extreme examples Sunstein worries about. So while it's possible to polarize, present research suggests many audiences haven't. The readership of the most popular websites, *Yahoo! News* and *CNN.com*, is apt to be more ideologically diverse than our neighborhood or workplace.

While that's a good thing, it also seems like a bare minimum for what we might expect from a news site: showing us perspectives beyond the ones we can encounter from our friends, families, and neighbors. But political ideology is only one form of diversity. If Gentzkow and Shapiro's study measured geographic isolation on the web, we would see vastly higher values. As we saw in chapter 2, most readers of the *Times of India* are in India (or are Indians living in the diaspora), whereas most readers of the *Globe* and *Mail* are Canadian. We may be reading news that's read by members of multiple political parties, but are we reading news that's read by people outside our home country?

Sunstein warns that we may be pushed to more extreme positions by spending too much time reading people who share our points of view and by deliberating with those who are like-minded. If we start considering our online behavior less in terms of the left/right domestic political spectrum and more in terms of an us/them view of the world, we may find we have a problem. Like broadcast media, curated media are far from balanced in offering a picture of the world. We see some countries more clearly than others. And our own desire to seek out and choose media is also affected by homophily.

If being surrounded by conservatives can convince you that cutting taxes will lead to a balanced budget (or surrounding yourself with liberals can convince you that deficits are irrelevant), how much influence does being surrounded by fellow Americans, Canadians, Chinese, or Danes have? The effect may not be the development of raw nationalism; it's likely to be a more subtle shaping of our worldview, suggesting that the issues most important to our neighbors are the most relevant international issues. We experience confirmation bias, the belief that an issue is important because our neighbors are convinced of its importance and confirm its significance. Other key issues and topics elude us, but the availability heuristic leads us to conclude that the topics we're hearing are the ones we need to know about. And since news has social currency, we benefit from spreading and talking about the news our friends care about and are interested in.

Are we more likely to see the world in terms of "us versus them" now that our picture of the world comes not just from newspapers but also through search engines? Given increased choice over what information we encounter, are we more likely to choose local perspectives over international ones? A recent lunch at Google suggests to me that this may indeed be the case.

Many wonderful things await visitors to Google's campus in Mountain View—art-filled, elegant buildings, open Wi-Fi networks, a wide array of free beverages. But the best part of a visit, in my opinion, is lunch. Many of Google's on-campus restaurants feature salad bars staffed by professional chefs. Once you've selected your ingredients, Google's salad chefs ladle your chosen dressing onto the salad, shake it between two bowls, and present it to you on a plate.

On my first campus visit, I assumed that the salad chefs were another manifestation of Google's obsession with efficiency, like offering employees dry cleaning and oil changes at the office. But something more complicated and subtle is going on. Salad dressing provides the easiest way to turn a healthful salad into a high-calorie meal. The second easiest is to pile on too much protein. Google's salad chefs control the amount of dressing and meat, ensuring you don't overindulge. And while you select the vegetables and topping for your bowl, the position of the ingredients on the salad bar is also designed to encourage moderation. High-fat ingredients like black olives and feta cheese are an awkward reach away, while raw veggies fill the front row. I shared my observation with an old friend, now a director of a division at Google, and she declared, "It's social engineering through salad."

Google's salad bar doesn't prevent you from serving yourself a dietary monstrosity laden with bleu cheese and bacon. It just decreases the chances that you'll do so accidentally, nudging you toward healthier eating choices. In that sense, it's much like the front page of a traditional paper newspaper. A daily newspaper's front page is laid out with a mix of local, national, and international stories. Often, the bottom of the front page will showcase a feature story buried deep within the paper, which a casual reader might otherwise miss. Major stories are presented with between 200 and 400 words of text, enough to capture a reader's interest and draw her off the front page and into the paper. The front page of the *New York Times* features roughly twenty "links" to stories deeper in the paper, avenues to begin exploring the content inside.

If the paper *New York Times* is Google's gently persuasive salad bar, the online version of the *New York Times* is a Las Vegas casino buffet. I counted over three hundred links to stories, sections, and other content pages in a recent analysis of the *Times* homepage. While there are vastly more links, there's vastly less to tell you what to follow: ten to twenty-six words associated with a story, on average. The paper *New York Times* is built to encourage serendipity. It's designed to help you stumble upon a story you might not have expected. And it shows us the curator's agenda, her sense that an international story is so important that it should occupy valuable front-page real estate. The online *Times* favors choice. It trusts us to know what we're looking for and to choose the news that interests us. And even though a growing number of users find their way to the newspaper through search engines, the home page still matters; the paper's assistant managing editor reports that 50–60 percent of people who visit the newspaper's website start at the front page.[27]

Can we choose our news wisely without any curatorial assistance? That's the danger of the search paradigm—we may choose what we want, not what we need. We may miss a story that's important for a large number of people, information necessary for us to be informed as local or global citizens. There's more choice, but also more responsibility. (On the *Times*'s current online home page, you can even choose between a US and a global edition, if you're worried your selections will be too global.)

Search increases choice at the expense of serendipity, the experience of discovering beneficial stories that we weren't intentionally seeking. When we can easily choose the news that interests us, we may miss stories that didn't appear interesting but that help us make unexpected and useful connections.

It's not yet clear whether the next paradigm shift for news, the social shift, makes us more or less likely to stumble on the unexpected and beneficial.

The Social Newspaper

In 2008 Lauren Wolfe, a young political activist, told a journalist about her news reading habits: "There are lots of times where I'll read an interesting story online and send the URL to 10 friends. . . .

I'd rather read an e-mail from a friend with an attached story than search through a newspaper to find the story."[28]

Wolfe's role as a disseminator of content is not unusual. A Harris Interactive study in 2006 found that 59 percent of American adults frequently or very frequently forward information found on the Internet to colleagues, peers, family, and friends.[29] The exchange of news stories through email and Facebook suggests a future where people don't make a decision to read the news. Instead, they simply encounter the news that their friends choose to amplify. In an age when many people are constantly connected via mobile phones, social networks, and email, word of mouth apparently delivers information to us all the time. It's not hard to conclude, as one college student told the media researcher Jane Buckingham, "If the news is that important, it will find me."[30]

She may be right, though the rise of social media as a pathway toward encountering news can be hard to see at first.

In an analysis of traffic to the twenty-five most prominent news sites for US audiences, the Project for Excellence in Journalism found that 60 percent of the traffic came directly to the website, not via a search engine query or link from another website. (The PEJ study considers only online media; the majority of people still get their news via television, radio, and newspaper, where a curated model prevails.) Search engines, particularly Google, direct roughly 30 percent of the traffic to news sites. And social media sites like Facebook drive less than 8 percent of traffic to the *Huffington Post*, the site receiving the most traffic from social media in the study. Other social media sites, including Twitter, drive far less traffic.[31]

These findings would seem to downplay the importance of social media. But it's worth taking a close look at the 60 percent of traffic that comes directly to a website. Some comes to the site's front page. But much of the traffic comes to individual URLs on the website. It's safe to assume that few people type in a URL as complex as http://globalvoicesonline.org/2012/10/12/

czech-republic-prednadrazi-struggle-continues/, so those visits likely result from links a visitor clicks from an email or an
instant message. Alexis Madrigal, a senior editor for the *Atlantic*,
calls these visits "dark social"—email and instant messaging are
clearly social behaviors, and they direct almost as much traffic
to the *Atlantic*'s website as search. Combine dark social—links
shared by email and instant messaging—with referrals from Facebook and Twitter, and social media are the dominant drivers of
traffic to the *Atlantic*.[32]

The *Atlantic* may be an outlier, but, as a whole, traffic from
social media is on the rise and traffic from search is falling. Facebook reports that traffic from its network to the average media
site doubled between 2010 and 2011. The rise was sharper at
some newspapers, with the *Washington Post* reporting nearly
three times as much traffic from Facebook as it experienced a year
before. Facebook now drives more traffic to the *Sporting News*, a
large US sports publisher, than any other website.[33] Tanya Cordrey,
director of digital development for the *Guardian*, partnered with
Facebook to create an app to feature *Guardian* content. In February 2012, traffic to the *Guardian* from Facebook crested at over
30 percent of referrers, surpassing traffic from search. Cordrey
describes this as "a seismic shift in our referral traffic," important
particularly because it's bringing younger readers to the *Guardian*'s site.[34]

As social media become a more powerful directors of attention, we are encountering less media through professional curators or through our own interest-based searches. In giving so
much responsibility to our friends to shape what we know of the
world, we need to consider the limitations of social discovery
rather than just celebrating its novelty.

Internet users now spend more than one of five minutes online
on social media sites like Facebook (22.7 percent of online time
for Americans in 2010, up from 15.8 percent a year before; 21.9
percent for Australians in 2010, up from 16.6 percent).[35] But this

type of social filtering is emerging throughout the web, not just on sites where friends post photos and updates.

Foursquare, a site that encourages users to "check in" at real-world locations like bars and coffee shops, now recommends places you might want to visit, based on locations favored by your friends and by people who frequent the same establishments you do. The music streaming service Spotify lets you share the music you're listening to with your friends and follow their recommendations for what to hear. Microsoft's search engine Bing integrates Facebook into the search experience, revealing what search results your friends have liked, and letting you post potential purchases to Facebook so your friends can "help you decide." Behind these tools is the idea that "millennials" (a generation of people born in the 1980s and 1990s) continually seek input from their friends and social networks in making decisions. *AdAge* reports that 68 percent of millennials consult their social network before making a "major" purchasing decision, which can be as minor as choosing what restaurant to eat at.[36]

The activist and author Eli Pariser is worried that this dependence on social networks shapes what we know about the world. His book *The Filter Bubble* begins with a story about his attempts to expand the kind of information he was encountering through social media: "Politically, I lean to the left, but I like to hear what conservatives are thinking, and I've gone out of my way to befriend a few and add them as Facebook connections. I wanted to see what links they'd post, read their comments, and learn a bit from them. But their links never turned up in my Top News feed." Pariser's conservative friends were not showing up because Facebook's EdgeRank algorithm prioritizes what's shown on your Top News page. The news you see is a product of time, the type of update (you're more likely to see a wall post than a comment), and your "affinity" with the person.[37]

This last factor is based on how often you visit that person's page or send him messages. You're more likely to receive news from someone you message with every day than from an old friend from high school whose links you seldom read. Pariser had a low affinity for the conservatives he'd added to his circle of friends: "Facebook was apparently doing the math and noticing that I was still clicking my progressive friends' links more than my conservative friends'. . . . So no conservative links for me."[38]

The rise of personalization technologies like EdgeRank, Pariser worries, will reduce our opportunities for serendipity and create a narrower world than we expect or hope to encounter. He's especially concerned about the invisible influences of personalization on tools like search, where we expect to see the same results for a search as everyone else but may actually be seeing content tailored by an algorithm specifically for us. His worries may be premature. Most of these services can be turned off, and Google engineers have responded to his book by suggesting that the effects he documents are generally less dramatic than the examples he cites.[39]

Pariser's concerns about social media isolation are a subset of broader concerns about ideological segregation. In his book *The Big Sort*, the journalist Bill Bishop, relying on research by Bob Cushing, a demographer, argues that Americans have physically relocated to communities where their neighbors are likely to share their ideology. Our communities are so highly segregated, Bishop observes, that many of the most powerful marketing techniques rely on the fact that individual demographics and psychological preferences are highly predictable on the basis of our postal codes.[40]

Relocating to a community of the like-minded entails selling our houses and packing boxes. Surrounding ourselves with the like-minded online requires a few mouse clicks, and, as Pariser points out, we've likely already done so. When you sign up for Facebook, the service first asks to search your email inbox and connect you to Facebook friends you're in email contact with. Next, it asks for your employer and for your high school and college information, including years of graduation, which it uses to introduce you to co-workers and classmates. Data from the Pew Internet and American Life Project suggest that most of our friends on Facebook come from these off-line associations: 22 percent are high school friends, 20 percent immediate and extended family, 10 percent co-workers, and 9 percent college friends. Only 7 percent of Facebook friends in the Pew study were "online-only" relationships; 93 percent were people whom a Facebook user had met off-line.[41]

Given what we know about homophily and given Bishop's observations about our tendency to segregate in the real world, it's fair to assume that our online friends aren't as diverse as the population of the nation we live in, and certainly not as diverse as the world as a whole. As social media become increasingly important as tools for discovery, it seems plausible that we're getting a less diverse view of the world than we might have encountered in the days of curated media, when professional editors endeavored to present a balanced news diet for us and our neighbors. Our eyes may glaze over when the *Guardian* runs updates on Paraguay, but in an age of social recommendation, we might not ever hear about the country unless we have Paraguayans in our social circle. Our filter bubbles are three-dimensional: they insulate us from content that is not just outside of our ideology but also outside of our orbits of geography and familiarity.

Shortly after Pariser published his book, I offered to Cameron Marlow my fears that social media were making readers of news more parochial. Marlow is a scholar of social media who became Facebook's "in house sociologist" in 2007. He responded with a provocation: we may soon hit a point where we're more likely to get news from another part of the world via a friend on a social network than via broadcast media. On its face, Cameron's idea seems absurd. The BBC has correspondents in more than a hundred nations, reporting global news, while the average Facebook user has 130 friends, most located in the same country. But Cameron's argument has a subtle twist: the question is not whether news will be reported by social networks or broadcast media but whether we, individually, will pay attention to it.

Many years before he became my boss at MIT's Media Lab, Joi Ito wrote me an email asking for links to African newspapers and blogs. He was traveling to South Africa for the first time and felt underinformed about the continent from American and Japanese media. I sent him links to some top newspapers and a few dozen bloggers I followed closely. He wrote me back weeks later with a heartfelt and frustrated message: he was having a hard time following the sources I'd offered because he knew very few Africans and felt little personal connection to the events he was reading about. Much as he wanted information from Africa, this "caring problem" was making it hard for him to pay attention.

If you don't know any Zambians, it can be hard to pay to attention events in Lusaka, the nation's capital. If a friend—perhaps one who has visited the country or befriended someone there—starts paying attention to news from Zambia, it sends a signal that stories from Zambia are important, at least to our local micropublic of Facebook friends. We might pay attention to the story as a way of showing our friend that we care too.

Or we might discover that we actually do care. In 1944, the communications researcher Paul Lazarsfeld proposed that media were less influential over public opinion than a "two-step flow of communication"—information that flows from media to an influential friend, and then from that friend to her friends. This idea of "opinion leaders" has been embraced by some sociologists and challenged by others, who argue that media have more direct influence than Lazarsfeld accounted for. But it's certainly possible that sharing news through social media signals the importance of a topic to our friends and encourages them to follow along.

My completely nonscientific, anecdotal experience of following news during the Arab Spring suggests that personal connection correlates to interest in news. Events in early 2011 rapidly reached a level of complexity that made it difficult to follow what was happening in all of the nations experiencing protests. At one point the *Guardian* published an online timeline that provided a simultaneous overview of events in seventeen nations.[42] (That timeline covered only the Middle East and North Africa; protests in Gabon, Sudan, Pakistan, and elsewhere inspired by the Arab Spring were not included.) Paying close attention to events in every country was a herculean task. Like most people, I watched and read what I was able, resigned to the reality that I was missing key events.

I recently looked back at my Twitter feed and blog posts to see what stories about the Arab Spring I shared with my readers. Stories about Tunisia, Egypt, and Bahrain dominate my feed, along with a few stories about Syria, but almost none about Libya. The imbalance is easily explained. My closest colleagues in the region are from Tunisia and Bahrain, and my interest in the events in their countries was deeply personal. I've never been to Syria, but the partner of a close colleague is Syrian, and her heavy tweeting about the situation there captured my attention. My lack of connection to Libya led to a corresponding lack of interest, which led to a lack of information. Tunisia and Bahrain, of course, are not inherently more worthy of attention than the conflicts in Syria and Libya, where far more individuals have lost their lives, but my experience suggests that while personal connections may not produce equitable results, their effects are powerful.

Whether Cameron's proposition proves to be correct depends heavily on two factors: What do we pay attention to in broadcast media, and how broad is our network of friends? The decreasing coverage of international stories in the American press suggests that publishers believe we don't pay much attention to international news. Homophily effects suggest that we're unlikely to know many Syrians unless we've lived, worked, or studied in Syria. But both those factors are more complicated than they look at first glance.

In 2006, I conducted an experiment where I tracked every story published on the BBC and *New York Times* websites, and used a blog search engine to see whether the stories were "amplified" by bloggers.[43] Many international stories went unamplified, but so did most local news. The stories that got circulated were ones that fit into larger, ongoing narratives: the political battle between left and right in the United States and the UK, the spread of terrorism around the globe, and the march of scientific and technological progress. I also found that "news you can use" stories—drink red wine and avoid cancer!—were widely spread. A study conducted by Jonah Berger and Katherine Milkman looked at a different form of amplification: the *New York Times* "most e-mailed" list. Stories that inspired people and evoked a sense of awe were forwarded, whether they were domestic or international, the study reported.[44]

It's not that we filter out international news; it's that we tend to filter out news that doesn't connect to our lives and our interests. For a story to be amplified, the connection can be quite loose. The strong US demand for news about the Arab Spring suggests interest in an inspiring story where familiar social media technologies appear to have played a role. Whether or not Facebook really helped organize protests in Cairo, the connection to a fascinating and familiar technology may have encouraged more Americans to follow the story.

Does Facebook raise the chances that you know someone in Syria? Paul Butler, an intern at Facebook, created a striking image in December 2010 that suggests it might: he plotted arcs between pairs of Facebook friends and produced a map that looks strikingly like an airline route map, or a map of the earth at night, showing users in almost every corner of the world.[45] A few months later, Marlow and colleagues released a study of friendships on Facebook that demonstrated that as many as 15 percent of relationships on Facebook cross an international border.[46] Given what we know about homophily, that's a strikingly high number.

The 2011 Facebook paper showed a matrix of friendships between people in different nations. Nations that are close together geographically, or that share historical connections through colonialism, tend to have more ties on Facebook. One explanation for this is language: if you have a common language, you can more easily maintain an online friendship. Speak a language that's not widely spoken in other countries, like Turkish, and you're likely to have fewer friends outside the country than if you live in a country like Jordan and speak Arabic. Smaller nations seem to have more international friendships than large ones.

I worked with Marlow and his colleague Johan Ugander to try to understand these differences. We found evidence of a connection between the mobility of individuals and the frequency of external friendships. (A subsequent visualization built by Mia Newman and Stamen Design helps confirm this, showing strong ties between the United States and Mexico, Brazil and Japan, and other pairs of countries linked by migration and by Facebook friendships.)[47]

Countries with high levels of migration, like UAE and Qatar, had highly international patterns of friendship, while less mobile societies like Nigeria were less connected. Facebook's data don't consider the country a user says he's from; they look at what country she logs in from. This means that strong ties of friendship between Pakistan and UAE probably don't represent burgeoning friendships between Emiratis and guest workers, but Pakistanis in Dubai keeping in touch with friends at home. This suggests another reading of Butler's elegant map of Facebook friendships, as a picture of global migration rather than a map of cross-border connection.

The data Facebook released considered only nations where the network has a strong foothold. If we could look at nations that are just starting to use the social network, we would see that countries where Facebook had just caught on had a higher percentage of international friends, while those on it for a long time had much lower percentages. This isn't evidence that Facebook is making people more parochial. It's evidence that the first people to use a tool like Facebook tend to have strong international connections.

Think of it this way: if you're the first person in Palau to sign up for Facebook, all your friends—by definition—will be outside Palau. You've probably discovered Facebook because you travel outside your country, or because you know people from other countries who are using the tool. As you introduce your friends in Palau to Facebook, your online social network begins to resemble more closely your off-line network—less international and more domestic. And the people who join later can look for friends in their own community, not unfamiliar people from around the world. Soon the people using Facebook aren't the most cosmopolitan Palauans; they're Palauans of all stripes. But they're just Palauans.

This phenomenon might explain some of the cosmopolitan enthusiasm about early social media projects like The Well or Usenet. It's possible that interactions in those networks were significantly less parochial than interactions on Facebook are today. The people attracted to online communication may already have had many connections to people in other countries, since online communication was and is an inexpensive way to stay in touch. And the structure of these networks, based on topical discussions of specific interests rather than on remaining connected to friends, may have decreased the tendency for online and off-line networks to merge. Before the rise of social networks designed to help you maintain existing off-line ties, the illusion of online cosmopolitanism was probably less illusory.

It's worth remembering that internationalism is only one possible manifestation of diversity. The Internet community scholar Judith Donath reports that she knows far more about the daily lives of her high school classmates in the age of Facebook than she did in the Usenet era. "Conversations on Usenet may have involved more nations, but the people talking had a great deal in common in terms of education level and occupation—it was all researchers and grad students in technical fields," she explains. "On Facebook, I'm in touch with people from a much broader socioeconomic range, because we went to school together."[48] That kind of diversity may help with the problems Cass Sunstein and Bill Bishop worry about. If we're connected to both a left-wing union activist and a right-wing evangelical because they were both in our high school class, Facebook might be a powerful tool for exposing us to diverse points of view. Or, if Eli Pariser is right, we may hear mostly from high school friends with consonant interests and opinions. In either case, unless you went to a very exciting high school, those viewpoints probably are not overwhelmingly international.

The most interesting piece of the conversation about Facebook cosmopolitanism came from an observation made by Johan Ugander, a Swedish American scientist who coauthored the paper on international ties on Facebook. "People like me have strong ties to more than one country—we're naturally going to have lots of international ties, and we're not typical of most people in a country." If the average American is connected to thirteen international Facebook friends, what we're actually saying is that 90 percent of Americans have far fewer, and 10 percent have significantly more international friends. Nations aren't cosmopolitan; people are.

People like Ugander who've lived their lives in different corners of the world are likely the key if we want social media to give us a broad view of the world and help us care about people we don't otherwise know. With a Swedish citizen in my network of friends, I'm likely to be exposed to news and perspective I otherwise would have missed. Whether that exposure turns into interest and attention is a function of my receptivity and Johan's ability to provide context around the news he's sharing.

Whether Cameron's prediction that we'll receive more international news through Facebook than through newspapers is right or wrong is ultimately irrelevant. I'm interested in finding ways to broaden my picture of the world and helping people who want to do the same. To encounter that wider world, we need to think about changing our media and broadening our circles of friends. We need to look at the media systems we've built, over hundreds of years in the case of newspapers and a dozen or so years in the case of social media, and ask whether they're working the way we need them to in a connected age. If they're not, we need to rewire.

REWIRE

Chapter 4

GLOBAL VOICES

We believe in free speech: in protecting the right to speak—
and the right to listen. We believe in universal access to the
tools of speech.

To that end, we seek to enable everyone who wants to
speak to have the means to speak—and everyone who wants
to hear that speech, the means to listen to it.

Thanks to new tools, speech need no longer be controlled
by those who own the means of publishing and distribution,
or by governments that would restrict thought and communication. Now, anyone can wield the power of the press. Everyone can tell their stories to the world.

We seek to build bridges across the gulfs that divide people, so as to understand each other more fully. We seek to
work together more effectively, and act more powerfully.

We believe in the power of direct connection. The bond
between individuals from different worlds is personal, political and powerful. We believe conversation across boundaries
is essential to a future that is free, fair, prosperous and sus-
tainable—for all citizens of this planet.

While we continue to work and speak as individuals, we
also seek to identify and promote our shared interests and

goals. We pledge to respect, assist, teach, learn from, and listen to one other.

We are Global Voices.[1]

AT 215 WORDS, THE MANIFESTO FOR THE GLOBAL VOICES project is shorter and marginally more readable than most manifestos, perhaps the most awkward of literary forms. It was produced in December 2004, near the close of a conference hosted by the Berkman Center for Internet and Society called "Votes, Bits and Bytes," a three-day conversation about how the Internet was changing the political process in the United States and around the world.

Berkman's annual conferences tend to reflect the conversations that are taking place in academic communities about the future of the Internet. In 2004, many of those conversations were about US electoral politics. Governor Howard Dean, of Vermont, had run unsuccessfully for the Democratic nomination for the US presidency, using the Internet to raise money, solicit ideas, and organize volunteers. Although he was trounced in the polls, Dean's use of the Internet served as a trial run for techniques used by Barack Obama in 2008. The idea that the Internet could be used to organize rallies and fund-raisers inspired a wave of thinking about how distributed participation might change politics and governance in general.

Two participants at the 2004 conference had written influential essays inspired, in part, by the Dean campaign. Joi Ito's "Emergent Democracy" explored the idea that groups of people could cooperate online to solve complex problems, anticipating Clay Shirky's writing on "organizing without organizations" and phenomena like crowdsourcing. Jim Moore's "The Second Superpower Rears Its Beautiful Head" built on the idea that Internet-enabled public opinion represents a check to American political power, and celebrated the possibility that bottom-up deliberative democracy will give citizens a voice in international institutions like the UN.

While I found these essays inspiring and challenging, they seemed limited by their focus on the developed world and specifically on the United States. Like Moore, I was a fellow at the Berkman Center and had come under the spell of Dave Winer, a brilliant if sometimes crotchety software developer who'd gotten the Center hooked on blogging. But I worried that blogs weren't giving voice to the voiceless, as the Deaniacs hoped, as much as they were providing another space for well-educated and well-connected Westerners to share their thoughts and opinions. I wondered whether we could test Moore's and Ito's rhetoric and use the Internet to make political dialogues more globally inclusive or, as I argued in a response to Moore's essay, whether we were ready to make room for the so-called "third world" in the "second superpower."

My partner in this experiment was Rebecca MacKinnon, a journalist who left her job with CNN after serving as its bureau chief of Beijing and Tokyo. She exchanged that lucrative and prestigious position for badly paid research fellowships at Harvard that enabled her to experiment with blogging and to research online citizen media because she was frustrated with CNN's waning commitment to international news. Rebecca is a respected China scholar, fluent in Mandarin, and deeply knowledgeable about Chinese politics. "The network kept telling me that my expertise was getting in the way. They wanted me to cover the region less as an expert and more as a tourist."

Rebecca's first few months as a visiting fellow at Harvard's Kennedy School of Government—before she decided to leave her CNN job—were spent running a blog about North Korea, which she covered as part of her CNN beat. North Korea is difficult to cover in a "normal" journalistic way because it is largely closed to the Western media. After coming to the Berkman Center, she began following and writing about the first bloggers who were starting to rise to prominence in China by late 2004. I was working with friends at AllAfrica.com, an online publisher of African newspapers, to curate BlogAfrica, a collection of blogs by Africans and travelers to the continent. As our contribution to the Berkman conference, Rebecca and I designed a "global bloggers summit" to run on the final day of the conference.

We invited some of the world's most visible and prominent bloggers to join us, stars of the nascent online world like Omar and Muhammed Fadhil of *Iraq the Model*, a relentlessly optimistic account of post-invasion Iraq; Hossein Derakhshan, an Iranian-Canadian whose Farsi-language guide to blogging had helped launch an Iranian blogging movement; and Jeff Ooi, whose passionate political blogging in Malaysia led to his election to parliament. But we were also joined by people we'd never heard of, like David Sasaki, a skinny, freckled Californian with an encyclopedic knowledge of the Latin American blogosphere, who'd flown on a red-eye from San Diego to Boston and slept on the floor of Logan Airport so that he could join our conversation.

Rebecca and I hoped that we would find some shared concerns and issues that united this disparate collection of bloggers. It was beyond our wildest hopes that this group would find common cause and begin to think of itself as a movement.

We quickly discovered that the writers we'd brought together had three things in common. They believed that their country was misunderstood and underrepresented on the global stage. This led them to write, as a way to represent their point of view as a counterbalance to how their country was often perceived.

One of the bloggers who inspired Rebecca and me to convene our meeting was the Bahraini author and activist Mahmood Al-Youssif, who explains his mission on his blog: "Now I try to dispel the image that Muslims and Arabs suffer from—mostly by our own doing I have to say—in the rest of the world. I am no missionary and don't want to be. I run several internet websites that are geared to do just that, create a better understanding that we're not all nuts hell-bent on world destruction."[2] Most of the people who joined us at Harvard had at least one myth to dispel, one perspective to counter with their writing.

Because bloggers felt that their voices needed to be heard, they were sensitive to the many ways they could be silenced, through censorship, intimidation, or sustained lack of public interest. Although online censorship was far less common in 2004 than it is today, the bloggers who joined us understood that there were no guarantees that speech would remain less constrained online than off-line. So it wasn't difficult for the group to agree on a collective focus on defending freedom of expression on the Internet.

The third common ground took more time to emerge. Not all people who are passionate about advocating for their country and ensuring that their voices can be heard are able to listen to others advocate for their own wants and needs. By the end of a long day, many of the bloggers we'd brought together realized that they felt an obligation to hear one another's stories and advocate on their behalf. Much as Kwame Appiah suggests cosmopolitans do, they saw themselves having obligations to one another, chief among them, the obligation to listen and try to understand. The language in the manifesto about "a right to listen" reflected the emergence of this common ground.

Building Global Voices

The Berkman conference had produced a document, but to make that manifesto real we created a website that would put those lofty ideas into action. Rebecca and I began posting excerpts from the blogs we followed, including the blogs of our manifesto coauthors. By the spring of 2005, the task had become unwieldy. We were each skimming hundreds of posts a day from different corners of the world. One day when both Rebecca and I were traveling, we asked the legal scholar Zephyr Teachout, then a fellow at Berkman, to take over the editorial reins for a day. "That was the most terrifying experience of my entire life," she reported back. "When you don't know anything about a country, it's almost impossible to know whether something on a blog is interesting or believable, or whether it's the work of a crazy person."

Clearly there were limits to a model where Cambridge-based intellectuals curated the world's blogs. Zephyr's discomfort helped us move to a model where responsibility for representing regions of the world rested with people from or living in those regions. We began to build a team of paid editors responsible for the daily work of collecting links to global blogs and for assembling teams of volunteers who would report on local issues by summarizing and excerpting important blog posts from their home countries. Reports offered translated excerpts of those posts and context on the events discussed, so that international audiences understood the issues and questions in play.

It quickly became obvious that the prominent, visible bloggers we'd built our conference around weren't the right people to take on this work. They had book contracts and speaking engagements. Bloggers who were deeply enmeshed in local politics would post the blow by blow of debates that made no sense to outsiders. Others shared posts only from those they agreed with, or from those they shared a language or ethnic background with.

Ultimately, we ended up relying on less visible bloggers who knew the local scenes well and could explain these issues to a global audience. Some, like Georgia Popplewell, a radio producer, who built a team to cover the Caribbean from her living room in Diego Martin, Trinidad, covered their home countries. Others, like David Sasaki, an American who built a Latin American team while teaching English in Monterrey, Mexico, were curious and well-informed interlopers. Virtually all had experience living or working in different parts of the world, like Lova Rakotomalala, who covered his native Madagascar while finishing a degree at Purdue University in Indiana. These digital cosmopolitans turned

Global Voices from an optimistic vision into a functional distributed newsroom.

Global Voices continues to work in this distributed way. Editors and volunteers worldwide find stories on blogs, photo- and video-sharing sites, and social media communities and share them on our website. But the project has grown arms and legs and now involves roughly nine hundred participants from more than a hundred countries. An advocacy arm, founded by the Tunisian human rights activist Sami ben Gharbia, documents threats to online freedom of speech and the arrest of bloggers and citizen journalists. Rising Voices, led by the Bolivian blogger Eddie Avila, supports online media from marginalized groups around the world. A massive team of translators produces versions of the Global Voices site in more than thirty languages, ranging from Arabic to Aymara, the indigenous language of Bolivia and parts of Chile and Peru. Roughly half a million people visit Global Voices sites a month, and content from our sites appears in media outlets around the world, including the *Economist*, the BBC, and the *New York Times*.

I'm more proud of Global Voices than of any other project I've helped build, but there's a real sense in which we failed. Rebecca and I saw Global Voices as a way to correct shortcomings in the professional media's coverage of the developing world: an over-reliance on uninformed "parachute" journalists, an emphasis on reporting natural disasters and violence at the expense of more complex and long-term stories, and an inability or unwillingness to feature the voices of people directly affected by events.

I had hoped Global Voices would influence agenda setting. In the simplest sense, I believed that by providing coverage of events that other media outlets missed, we would help challenge the imbalances in attention that Galtung and others had documented. Rebecca, more experienced in the ways of broadcast media, was less confident of our ability to sway news agendas. But both of us hoped that by offering a global perspective through the eyes of a specific individual on the ground, readers would have an easier time connecting with unfamiliar stories.

Instead, Global Voices has become a go-to source for information on the infrequent occasions that countries rarely in the news suddenly burst into the headlines. We'd tracked civil unrest in Tunisia from protests in Gafsa in 2008 through Sidi Bouzid in 2010 and found ourselves inundated with calls for assistance during the week that Tunisia's revolution dominated the news agenda. As Jennifer Preston of the *New York Times* wrote about our coverage of the Arab Spring, "When unrest stirs, bloggers are already in place."[3] That's true, and important, but in practice it means that Global Voices offers reporters a way to get quotes from countries experiencing sudden turmoil, rather than using us to find important unreported stories before they break.

The power of personal connection has proven to be both stronger and weaker than expected. We'd hoped that more bloggers featured on Global Voices would emerge as guides to their cultures, like Salam Pax, the Austrian-educated architect whose accounts of the US occupation of Iraq helped personalize the war for supporters and opponents of the invasion. We've seen few individuals emerge with a profile like Salam Pax's. Instead, citizen media reports are often eyewitness accounts, capturing attention not because of a relationship with the observer but because she was in the right (or, more often, wrong) place. Perhaps this is because few global events have captured America's attention like the Iraq War, but it seems that conflicts like the Syrian uprising could use a guide to explain events on the ground to a global audience.

Still, personal connection is the glue that has allowed Global Voices to thrive with little money, little central organization, and very little face-to-face contact. Aside from a biennial meeting that's part media conference and part global dance party, the connective tissue of Global Voices is a set of mailing lists, where apparently trivial updates, like birthday wishes and announcements of engagements and children's births, are far more common than heady discussions about the future of media. Global Voices survives because it's a social network, connecting people with common interests and purposes across lines of language, faith, and nation. What has resulted is a powerful team of people who support one another, in ways as mundane as providing a couch to sleep on and as profound as managing international campaigns to release fellow community members arrested by their governments for their online writings.

The successes and failures of Global Voices offer an object lesson in the challenges of using the Internet to encounter a wider picture of the world. Seven years of arguments over the day-to-day matters of running an international newsroom and accompanying nonprofit organization have proven an excellent reminder of the challenges inherent in moving from theory to practice. A project like Global Voices wasn't possible before the spread of the Internet and the rise of participatory media. But the fact that people are tweeting reports from demonstrations in Syria doesn't inexorably lead to media that are more fair, more just, or more inclusive than analog media. Even when it's possible, connecting the world is hard work.

Global Voices got several things right, rarely because we figured out what we were doing in advance. Instead, we tried and failed, and tried again, on the path toward solutions. Our basic model for sharing citizen media from around the world is a likely framework for any project that uses the Internet to expand worldviews. We curate, searching through the vastness of participatory content to find the bits that illuminate issues, concerns, and lives in other parts of the world. We translate, opening a conversation beyond its linguistic borders. And we contextualize, explaining what events mean to people on the ground and what they might mean to you. Whether you're trying to follow news from Thailand or to collaborate with an artist in Senegal, filtering, translating, and contextualizing are going to come into play.

One major shortcoming of Global Voices was a failure to understand principles of supply and demand. Like many well-intentioned reformers, I had assumed that the world wasn't paying enough attention to international news because there simply wasn't enough being reported. I thought that the problem was the financial shock the journalism industry was experiencing from the rise of the web, and that by supplying a set of novel perspectives and free, high-quality content, journalists would flock to our project and amplify stories to their audiences. When journalists didn't beat a path to our door, I began researching the dynamics of international news and coming to understand that foreign news bureaus weren't closing just because they were expensive to operate.

Inequities in media attention are in part a demand problem. If audiences aren't interested in Madagascar, the wealth of stories we provide from that strange and fascinating nation go unread unless we can help audiences see how strange and fascinating they are. As we figure out how Global Voices can fulfill its mission, we're thinking of our work—and particularly our curatorial work—in terms of building demand, in part by helping people find not just what they're interested in, and not what we believe they should be interested in, but what they're surprised and delighted to discover they're interested in.

Global Voices reports on coups and protests, but also on how families in different countries break the Ramadan fast. Our most popular posts often aren't breaking news stories, but stories on the intricacies of South Korean pop music or Nigerian film. Building demand means unpacking culture as well as news. Knowing that Mongolia has the fastest-growing economy in the world in 2011 is helpful if you're an adventurous investor, but discovering that Mongolians are dominating Japanese sumo and causing controversy in the process might be more helpful in developing your interest in Central Asia.

It's hard to solve a difficult problem through theory alone. If we want digital connection to increase human connection, we need to experiment. We must build things, test them, and learn from our failures. This section of the present book suggests three areas where the Internet as we know it needs rewiring: language, personal connection, and discovery. On the basis of the lessons learned from Global Voices, I offer three possible ideas to explore: transparent translation, bridge figures, and engineered serendipity. We're far from solving these vast problems, but we think we've found areas where we could all benefit from some attempted rewiring. What follows is an introduction to some of the efforts already underway, and an invitation to join me and my friends in the work we've been doing: rewiring the web for a wider world.

Chapter 5

FOUND IN TRANSLATION

IT'S A CLICHÉ TO OBSERVE THAT AMERICANS DON'T KNOW much about football. First, we tend to call it "soccer." Second, many Americans tune into the world's game only once every four years, noisily supporting our national team during the World Cup, and then turning attention back to college football, NASCAR, and other distinctly American pursuits. When we do tune in, it's expected that we'll need a refresher on the offsides rule and a reminder of some key terminology.

As viewers around the world tuned into a surprisingly close game between Brazil and North Korea at the start of the 2010 World Cup in South Africa, many caught a glimpse of a banner held by Brazilian fans that declared, "Cala Boca Galvão." Those who watched matches while trash-typing on Twitter saw the phrase repeated thousands of times by Brazilian fans. Four days into the monthlong tournament, Twitter listed "Cala Boca Galvão" as a trending topic, a word or phrase receiving an unusually high amount of attention from the service's global user base. Was this a message of support for the Brazilian team?

While Twitter told its users that "Cala Boca Galvão" was popular, it didn't help them figure out what the phrase meant. Fortunately, Brazilian users were willing to help out. The Galvão bird, they explained, was endangered, hunted nearly to extinction for its colorful feathers, which adorn the headdresses of the samba schools that dance in Carnival parades. The Galvão Institute was established to raise awareness of the bird's plight, and if other Twitter users tweeted the phrase "Cala Boca Galvão"—Save the Galvão Bird—a donation of $0.10 would be made to the cause. A slick, English-language video produced by the Galvão Institute and posted on YouTube provided background on the plight of the Galvão and urged participation with the tagline: "One second to tweet, one second to save a life."

Clearly the campaign was working. Not only did Twitter users rise to the defense of the Galvão bird, prominent celebrities took up the cause as well. The pop idol Lady Gaga was rumored to be releasing a new single, titled "Cala Boca Galvão," and dozens of YouTube versions of the song appeared. Many versions sounded like a reworking of Gaga's "Alejandro," though some, strangely, appeared to feature entirely different melodies. The Galvão Bird Foundation, a sister organization to the Galvão Institute, exposed a darker side of the issue with a revealing photo of the Argentine football coach Diego Maradona with a green feather sticking out of his nostril. Evidently, Galvao birds were also desirable for their hallucinogenic properties.

For those who hadn't already typed the phrase into a site like Google Translate, the *New York Times* blew the joke with a June 15, 2010, story that revealed the simple truth: "Cala Boca, Galvão" translates as "Shut up, Galvão." Carlos Eduardo dos Santos Galvão Bueno is the primary football commentator for Rede Globo, the television network carrying World Cup games in Brazil. His constant cliché-ridden patter alienated many Brazilian fans who wished he'd just shut up. The phrase caught on as thousands of Brazilian fans watched the opening matches on Rede Globo and began venting their frustration. Once the phrase appeared on Twitter's trending topics, it became a game to maintain the phrase's popularity. By encouraging unsuspecting, well-meaning non-Brazilians to spread the phrase, it turned into a vast joke wired Brazilians played on the rest of the world.

There are a number of possible lessons we can take from the Cala Boca Galvão story. First, a lot of Brazilians are on Twitter— over 5 million, 11 percent of the country's online population, when this story took place, and many more now. Second, at least some of those Brazilian netizens have a wicked sense of humor. A later iteration of the Cala Boca meme urged people to save the Geisy Arruda whale, an unkind reference to a curvaceous Brazilian woman expelled from a university in São Paolo for wearing a miniskirt. Most important for our discussion here, however, is the lesson that linguistic difference persists in the face of globalization, and that this difference is a barrier to connection and understanding.

A connected world is a polyglot world. As we gain access to the thoughts, feelings, and opinions of people around the world, our potential for knowledge and understanding expands. But so does our capacity to misunderstand. As we become more connected, we're able to comprehend a smaller and smaller fraction of the conversations we encounter without help and interpretation.

A Lingua Franca?

Conventional wisdom suggests that English is becoming "the world's second language," a lingua franca that many forward-looking organizations are adopting as a working language. Optimists about the spread of English as a global second language suggest it will enable collaboration and ease problem solving without threatening the survival of mother tongues. Pointing to hundreds of thousands of Chinese children who learn English by shouting phrases back at teachers, the American entrepreneur Jay Walker offers the idea that English will be a language of economic opportunity for most speakers: they'll work and think in their mother tongue, but English will allow them to communicate, share, and transact.[1]

Cultural-preservation organizations like UNESCO aren't as confident of this vision. They warn that English may crowd out less widely spoken languages as it spreads around the world through television, music, and film. But something more subtle and complicated appears to be going on. While English may be emerging as a bridge language, a wave of media is being produced in other languages, in newspapers, on television, and on the Internet. As technologies make it easier for people to communicate to broad and narrow audiences in their native languages, we're discovering that linguistic difference is surprisingly persistent.

One way to consider the future of language in a connected world is to ask, "What percent of the Internet's content is written in English?"

Look online for an answer to that query—posed in English—and you're likely to encounter a website last updated in 2003, EnglishEnglish.com. The site's "English Facts and Figures" page asserts that "80% of home pages on the Web are in English, while the next greatest, German, has only 4.5% and Japanese 3.1%." The sources behind this assertion are unclear, but it's consistent with early research on linguistic diversity online. In 1997, Geoffrey Nunberg and Hinrich Schütze released a study estimating that 80 percent of the World Wide Web's content was in English. The Online Computer Library Center followed in 2003 with a study estimating that 72 percent of online content was in English.[2]

These early studies led researchers to suggest that English had a "head start" that other languages would find difficult to overcome. With such a large user base of English speakers online, many websites would publish content only in English, and web users would adapt to monolingualism by improving their language skills, which in turn would increase the incentive to publish in English. Neil Gandal of Tel Aviv University analyzed web use in Quebec, Canada, in 2001 and observed that native French speakers spent 66 percent of their online time on English-language websites. Furthermore, young Quebecois looked at more English content than their elders, suggesting that language barriers would be even less relevant for a future generation of web users.[3] Given that Francophone Quebecois were willing to read English content online, Gandal argued, website developers wouldn't bother to localize their content, leading to a future with more sites entirely in English.

Both the 70–80 percent English "fact" and the head start theory have lingered despite evidence that the linguistic shape of the World Wide Web has changed dramatically in the past ten years as it expanded both in scale and in the number of authors creating content. One reason the "fact" persists is that it's incredibly difficult to generate a believable estimate of language diversity online. Early studies tried to create a random sample of websites by choosing a selection of IP addresses, loading whatever page emerged, and using automated tools to determine what language it was written in. This method works poorly these days, when sites like Facebook, reached via a single IP address, include multilingual content generated by more than half a billion users. Newer methods rely on search engines to index the web, then attempt to estimate coverage of different languages on the basis of the comparative frequency of words in different languages.

Álvaro Blanco leads a team at FUNREDES (Foundation for Networks and Development), a Dominican Republic–based nonprofit organization focused on technology in the developing world, that has been researching linguistic diversity since 1996 by means of these new methods. Try your search query about English-language content online, posed in Spanish or most other Romance languages, and it's his research that usually tops the search results. His team searches for "word concepts" in different languages, counting the results for "Monday" versus "Lunes" (Spanish) versus "Lundi" (French). In 1996, his research estimated that 80 percent of the content online was in English. That percentage fell steadily through successive experiments until 2005, when he estimated that 45 percent of online content was in English.

While Blanco's research continues, he warns that search engines may no longer offer a representative sample of content online. "Twitter, Facebook, social networks—these are all difficult for search engines to index fully." Blanco estimates that search engines now index less than 30 percent of the visible web, and suggests that the indexed subset skews toward English-language sites, often because those sites are the most profitable places to sell advertising. "My personal opinion is that English now represents less than 40% of online content,"[4] Blanco offers, though he believes he'll need to refine his methodology to prove his hunch.

Statistics about Internet usage show much faster growth in countries where English is not the dominant language. In 1996, more than 80 percent of Internet users were native English speakers. By 2010, that percentage had dropped to 27.3 percent. While the number of English-speaking Internet users has almost trebled since 2000, twelve times as many people in China use the Internet now as in 1996. Growth is even more dramatic in the Arabic-speaking world, where twenty-five times as many people are online as in 1996.[5]

But that's not the most important shift. When Gandal predicted that Quebecois web users would get accustomed to using sites like Amazon.com in English, he didn't realize that most web users in 2010 would be creating content as well as consuming it. More than half of China's 450 million Internet users regularly use a social media platform, writing blog posts, posting updates on Renren (China's Facebook equivalent) or status messages to Sina Weibo, a microblogging site similar to Twitter. And the vast majority of those updates are in Chinese, not English.[6]

On a visit to Amman, Jordan, in July 2005, the high point for me was a leisurely dinner with a dozen Jordanian bloggers, whose websites I'd been following in the run-up to the trip. As we looked over the ancient stone houses of Jabal Amman from the terrace of the restaurant, our conversation bounced between English and Arabic. "You guys all speak Arabic as a first language—why do you all blog in English?" I asked. Ahmad Humeid, a talented designer and the proprietor of the *360° East* blog[7] explained, "I want my perspectives on Jordan to be read around the world, which means I need to write in English. Besides, the people who only read Arabic aren't reading blogs."

Seven years on, Ahmad still blogs in English, but many newer Middle Eastern bloggers write primarily in Arabic. For multilingual web users, there's a tipping point associated with language use. So long as most of your potential audience doesn't speak your language, it makes sense to write in a second, more globally popular language. But once your compatriots have joined you online, the equation shifts. If you want to reach your friends, you may write to them in one language. If you want to engage a wider audience, you may use another. Haitham Sabbah, a passionate Jordanian Palestinian activist who served as Middle East editor for Global Voices from 2005 to 2007, now writes in English to criticize American and Israeli policy in the Middle East and in Arabic to critique Arab leaders, making those criticisms more opaque to international audiences. English is for engaging with a wide audience, while Arabic is a private language for disagreements he has with fellow Arabs, which he wants to keep "within the family."

Gandal's Quebecois research subjects may have read a lot of English-language content, but that doesn't mean they preferred reading in a second language. While most of India's 50 million Internet users speak English, a survey by the Indian market research company JuxtConsult revealed that almost three-quarters prefer and seek out content in their first languages.[8] Cognizant of this preference, Google offers interfaces to its search engine in nine different Indian languages, and in over 120 languages in total. Given that 68 languages are spoken by at least 10 million speakers worldwide, other companies with global ambitions may be looking at developing Tagalog and Telugu interfaces in the near future.

When we began curating blog posts to publish on Global Voices, Rebecca and I realized we would need to address issues of language and translation. We hired editors fluent in French, Arabic, Russian, Chinese, and Spanish to translate conversations into English for publication on the site. In those early days, we never seriously considered publishing an edition other than in English, assuming that translating our work into other languages would be prohibitively expensive and that, since our community of editors and authors used English as a "working language," everyone could read and appreciate our output.

Less than a year after we started the project, Portnoy Zheng, a Taiwanese university student, launched a Chinese edition of the Global Voices site. Taking advantage of the fact that Global Voices publishes using a Creative Commons license, Zheng and friends began selecting stories from Global Voices that caught their attention and posting Chinese translations on his website. After Portnoy accepted our offer to turn his site into an official Chinese edition of our site, hosted on our servers, Rebecca and I were flooded with requests to build other-language editions of Global Voices.

Why does it make sense to produce Global Voices in Malagasy, a language rarely spoken outside Madagascar, a country where only 1.5 percent of the population has access to the Internet? Our Malagasy contributors were worried that their language wouldn't make the leap from the analog to the digital world. French, not Malagasy, is taught in schools, and French enjoys a higher prestige than Madagascar's indigenous language. Our contributors were willing to do the work to publish the edition and help preserve the language. Though they personally were trilingual, they wanted to share their work, and the broader coverage of Global Voices, with friends and family who weren't as comfortable reading English or French as they were.

Our Malagasy site is now read by a significant fraction of Madagascar's online community and has inspired a new humility on the part of our editorial team regarding the importance of language. Translators, responsible for making our content accessible in more than thirty languages, now outnumber writers of original content for the site, and those sites, collectively, receive as much traffic as our English-language site.

In 2010, members of our community asked for an additional change to Global Voices: they wanted to publish original content in French, Spanish, and other languages besides English. This presents a challenge for our editorial team. While virtually everyone involved with the project speaks multiple languages, it's hard for our editor in chief to take responsibility for posts in languages she does not speak.[9] After a long debate, we reached a consensus, and now our multilingual newsroom translates stories written in over a dozen languages into English. This leads to uncomfortable moments: I sometimes glance at our servers and discover our most popular (often our most controversial) story is in a language I don't read well, and I find myself waiting for our French-to-English translators to catch up so that I can understand what our team is publishing. But it's clearly been the correct step to take. Our coverage of Francophone Africa is much stronger than in past years, because authors who can write easily in French can now compose in that language, then rely on a community of translators to make their work accessible in English.

Language Is a Tool

To understand why it's so important for our volunteers to write in their native languages, and why most web users will create more and more content in their own languages, it's useful to consider language as a technology, a tool humans have created that can be applied to solve a wide range of problems.[10] When we begin using any new tool—a screwdriver, a car, a computer—we tend to be acutely aware of the tool itself, the challenges of using it, its limitations and potentials. As we become increasingly familiar with the tool, it becomes increasingly transparent to us.

In "The Disappearance of Technology," the information scholar Chip Bruce observes that, at a high degree of fluency, tools simply become invisible: "We might say, 'I talked to my friend today,' without feeling any need to mention that the telephone was a necessary tool for that conversation to occur."[11] (Or, for that matter, language: "I talked to my friend today using words, in English.") That invisibility is a benefit. We use tools more effectively when we think not about the instruments at hand but rather about the task we're trying to accomplish. But that invisibility makes it easy to forget the biases associated with the tool. Certain places are easier to get to on foot than by car, and certain information is easier to find in a library than online. As one of the most pervasive and powerful tools we use, language biases what we encounter, and fail to encounter, every day.

For those who don't speak English as a native language, language biases are all too clear in online spaces. The task of learning to use a new tool is complicated by the fact that the interface and instructions are in an unfamiliar language. Achieving fluency—the invisibility of the technology—takes longer, and the learning curve is steeper. Creating content online in a language like Hindi requires an author to install a new font and a keyboard driver that will allow an English-language keyboard to create the appropriate characters. It's so complex and awkward that many Hindi speakers use Quillpad, a piece of software that allows you to type Hindi words transliterated into English characters and have the results appear in Devanagari script. Given the barriers to creating content, the sharp rise in content created in languages like Hindi should hint at the importance readers and writers place on local languages.

Those of us who do speak English as our first language need

to consider transparency and biases in another way. It's easy to assume that the most important content will appear in the language we speak at some point. That's no longer a safe assumption. Each day the amount of information we could encounter via broadcast or online media increases while the percentage we can understand shrinks. The opposite is true for speakers of languages like Arabic, Chinese, and Hindi, whose representation online is growing.

Wikipedia, the remarkable collectively written encyclopedia, was a multilingual project almost from inception; German and Catalan editions of the encyclopedia were launched two months after the initial English-language launch of the project, in January 2001. Rather than create a master encyclopedia in one language and produce other editions through translation, early Wikipedians realized that collaborative encyclopedias needed to be written independently in different languages so that they could reflect local priorities.

An ecosystem has emerged in which many Wikipedias have a core of articles that exist in other languages and a large set of articles unique to that language. While both French and English Wikipedias feature long- and well-researched articles on Charles Darwin, the sociologist Paul-Henry Chombart de Lauwe (whom we'll encounter in chapter 7) merits an article only in the French Wikipedia. As we look for information outside the core subjects covered in many languages, monolingualism emerges as a barrier. A 2008 study of English, French, German, and Spanish Wikipedias suggests that the 2.4 million–article English-language Wikipedia had 350,000 articles covering the same topics as the 700,000–article French-language Wikipedia, which implies that half the French-language Wikipedia wasn't accessible to English speakers, and over five-sixths of the English-language Wikipedia was closed to Francophones.[12] A great deal of knowledge is inaccessible to people who speak only English or French.

The challenge of accessing information in languages we

don't speak can lead also to misunderstanding and misinterpreting what we know. In January 2010, Google reported that its servers had come under sustained cyber attack by Chinese hackers in search of both corporate secrets and the personal email accounts of human rights activists. On February 18, 2010, the *New York Times* broke a story by John Markoff and David Barboza that traced the attacks to two Chinese universities, the elite Shanghai Jiaotong University and the much lesser known Lanxiang Vocational School. The *Times* report characterized Lanxiang as a military-connected technical college and reported that the attackers had studied with a specific Ukrainian professor of computer science at the university. More than eight hundred English-language news outlets printed some version of the *Times* story, though a study conducted by Jonathan Stray for the Nieman Journalism Lab found that only thirteen of those accounts included original reporting.

The story caught the attention of Chinese audiences, and though Chinese journalists were not surprised that Shanghai Jiaotong University might be implicated, the inclusion of Lanxiang Vocational School raised some eyebrows. The school advertises on late-night television commercials with the tagline "Want to learn to operate an earth extractor? Come to Lanxiang" and is known best for offering degrees in auto repair and truck driving. Reporters from the *Qilu Evening News*, a government-sponsored newspaper with a circulation of over a million copies, visited Lanxiang shortly after the *Times* story broke and reported that the university had no Ukrainian professors, that the military ties extended to Lanxiang graduates' repairing army trucks, and that computer classes at the school taught word processing and some basic image editing. Their story, which included slams at the *New York Times* for its credulity, ended with the observation that Chinese netizens were circulating a joke, "Want to learn to become a hacker? Come to Lanxiang in Shandong, China."

It's understandable that English-language news outlets weren't

able to travel to Lanxiang to verify the *Times* story, and understandable, if disturbing, that outlets reporting on China aren't
able to read reporting in major Chinese newspapers. But the *Qilu*
story was available in English within twenty-four hours of publication, posted on EastSouthWestNorth, a website run by the
widely respected Chinese-to-English translator Roland Soong.
While Soong's site is daily reading for many English speakers who
follow Chinese media, journalists covering the story missed the
Qilu story, suggesting that even when translations of key stories
in other languages exist, it's easy to miss them unless they're part
of our information discovery process, as visible in our inboxes or
search engines as domestic news sites.

The *New York Times* got the story wrong, presumably, because
its sources had inaccurate information. Other English-language
newspapers got the story wrong because they followed the *Times*,
but also because they couldn't, or didn't, read Chinese accounts
of the same events. We are still a long way from an Internet where
English-speaking reporters can triangulate between Chinese and
English language sources to understand Chinese events.

The Checkered History of Machine Translation

On January 7, 1954, a team from Georgetown University and
IBM held a demonstration of a remarkable new tool at IBM's
New York headquarters: a computer system that translated Rus-
sian sentences into English. As Robert Plumb reported in the
New York Times the following day,

> In the demonstration, a girl operator typed out on a keyboard
> the following Russian text in English characters: "Mi pyerye-
> dayem mislyi posryedstvom ryechi." The machine printed a
> translation almost simultaneously: "We transmit thoughts

by means of speech." The operator did not know Russian. Again she types out the meaningless (to her) Russian words: "Vyelyichyina ugla opryedyelyayetsya otnoshyenyiyem dlyini dugi k radyiusu." And the machine translated it as: "Magnitude of angle is determined by the relation of length of arc to radius."[13]

Although the Georgetown/IBM system had a vocabulary of only 250 words and knew only six grammatical rules, its success was a technical triumph, given that the computer system it ran on, the IBM 701, had total data storage of 36Kb and had to be programmed in assembly language[14] by the IBM systems programmer Peter Sheridan. Because programming the 701 was so difficult, Sheridan prototyped the software by writing a set of English-language instructions and giving them and a set of dictionary cards to non-Russian-speaking volunteers. Volunteers searched through the deck of cards to find the appropriate Russian word and the corresponding English word, and then worked through Sheridan's instructions to add or subtract stems from words or rearrange their order in the sentence.

If the scope of the 1954 demonstration was limited, translating sixty carefully selected sentences, the ambitions of those behind it were not. Professor Léon Dostert, who developed the language model Sheridan painstakingly programmed, noted that while it was not yet possible "to insert a Russian book at one end and come out with an English book at the other," in the future "five, perhaps three, years hence, interlingual meaning conversion by electronic process in important functional areas of several languages may well be an accomplished fact." Building these systems, Dostert suggested, would require a dictionary of 20,000 words and one hundred rules, essentially a scaling up of the prototype.[15]

Dostert's prediction sounds laughably optimistic in retrospect, but the system he contemplated was being designed to translate scientific journals, not Tolstoy or Pushkin. Dostert knew that dictionary-based translation systems have a great deal of difficulty with linguistic ambiguity and that natural human language is extremely ambiguous. Many languages feature homonyms, words with identical spelling but different meanings, or polysemy, where the same word can have related, but different, meanings depending on context: "Take note! I left a note for the trumpet player about the note she needs to play." More complicated phenomena like metaphor, allegory, or puns add layers of complexity to the task of translation; they make the process difficult to replicate by looking up words in a dictionary and ordering them into a grammatically correct sentence.

When a human translator decides how to translate the word "note," she reads and understands the sentence, then chooses an appropriate word in a target language on the basis of the context the word was used in. Most of the sentences tested in the 1954 demonstration were from physics and chemistry, both because the promise of the Georgetown/IBM system was the ability to translate scientific literature and because the context of scientific literature reduced the ambiguity around some of the terms used.

To solve the problems of context and to make it possible to translate "note" correctly, more modern translation systems throw out the dictionary and grammatical rules and work instead with statistics and probabilities. These systems are built around huge piles of text, called corpora. Most systems rely on two corpora. One is a collection of sentences in a target language, which allows programmers to develop a "language model." By analyzing this collection of sentences, the language model "knows" that the phrase "the blue car" is more common in English than "the car blue," and in choosing between those possible translation outputs, it can choose the grammatically correct one, not because it understands grammatical rules but because the correct one is the most common one. A second corpus collects sentences that have been translated by humans between a pair of languages to create a "translation model." The translation model tells us that "el coche azul" in Spanish is translated as "the blue car" quite frequently in English, though occasionally "the azure auto" might appear in a document. Translating a new document becomes a series of educated guesses, choosing the likely sentence equivalents through the translation model and ensuring they're grammatical and readable through the language model.

This method—statistical machine translation—was impossible before the late 1980s. Until then computers simply couldn't handle the huge sets of data needed to build workable language models. While it was challenging for the Georgetown/IBM system to maintain a 250-word dictionary, the corpus Google has released to the public as an English-language model consists of over 95 billion English sentences. Given the size of the data sets needed in order for this method to be effective, search engines have the upper hand in building them. Indexing the Internet offers a great opportunity to expand language models. But even Google is often constrained by the need to find reliable *parallel corpora*, sets of sentences that have been translated between one or more languages.

Parallel corpora are hard to find because high-quality human translation is (traditionally) very expensive. For these systems to be useful, they must be huge. The Linguistic Data Consortium's parallel corpus for English/Chinese translation includes 200 million words, far more words than exist in either language, because to be effective those words need to exist in many different contexts. Many corpora we might like to use—translations of Stephen King novels into dozens of languages, for instance—are off-limits because of copyright constraints. Looking for high-quality, open-licensed text, programmers often rely on corpora that are collections of government documents: official UN resolutions translated in the institution's six working languages; the European Parliament's proceedings, which include documents translated between the twenty-three official languages; Canadian government documents, which exist in English and French.

Because statistical machine translation is basically the process of selecting a likely translation from a set of examples, there's an odd implication from the origins of these systems: in translation, we may all sound a little like European parliamentarians. In practice, these systems tend to do better in translating formal documents than in translating slang and jargon-filled instant messages.

So why weren't American and European newspaper reporters and fact-checkers reading the *Qilu Evening News* through a mechanical translation system to get a fuller understanding of the Lanxiang Vocational School? In part, their decision might have been force of habit. For many years, mechanical translations were awkward and imperfect, and reporters may have developed a bias against using them. But the quality of machine translations between Chinese and English has increased dramatically over the past five years. Programmers evaluate the success of machine translation systems by comparing their output with outputs generated by professional human translators. From the comparison, they calculate scores like the Biligual Evaluation Understudy, or BLEU score, which measures whether a machine translation includes the same words, in the same order as a professional translation. When Google determines that a BLEU score for a new translation pair (English/Chinese, for instance) is high enough, it's released and included in the set of tools Google makes available for free at translate.google.com. Over six years, from 2006 to 2011, sixty language pairs crossed this threshold.

Newspaper reporters might be impressed by what they see, reading a machine translation of the *Qilu Evening News* story. I used Google to translate the story and got, in part, the following:

Zhou did not meet with the school office reporter, but on the phone to respond: "These reports are fabrications. A few days ago, a Chinese-speaking women to call for consultation in the name of recruitment cliches, and no Liangmingshen-fen We are mainly specialized vehicle maintenance, vehicle maintenance down there are some students who do join the army after the mechanical maintenance activities. said Professor Ukraine to teach here, is off the mark, the school is not foreign teachers, we do not have to use the teacher's qualifications. Besides, we are not refusing to answer whether the Ukrainian foreign teachers, but she did not ask ah. "

The English text that emerges is somewhat comprehensible, but far from easy to read. It's unlikely that anyone would mistake this passage for one written by a native English speaker. An intrepid reporter might find the *Qilu* story in translation and use it to enhance her story. But it's unlikely that any English speaker would try reading the *Qilu Evening News* each day through machine translation as a way of keeping up on events. And, if she did, there's a strong chance she could misunderstand the article in question.

When IBM and Georgetown began translating Russian sentences, the goal was to create a system that could automate some of the translation of scientific journal articles, recognizing that those translations would need to be hand-polished before delivery to American scholars. As the program struggled to make gains in the early 1970s, government funders backed away from automated machine translation and focused instead on building tools that could help make human translators more efficient—software like "translation memories," which store how a translator has interpreted a complex phrase and make that translation available to him and to other translators he works with. The goal for US government systems became making human translators more efficient, rather than perfecting automated translation.

The gaps between Soviet and US science are no longer as politically important as they were in the 1950s. As we've moved beyond the Cold War into a complex, multipolar world, the US government audience for international media is now the intelligence community, notably the Open Source Center, a section of the CIA that tries to understand global events by reading local newspapers published in Pashtun or Azeri (among others). Newspapers like the Baku *Xalq Qэzeti* are translated by human translators for the benefit of CIA analysts. Their work is available to the general public as well . . . sort of. The US Department of Commerce packages the unclassified work produced by translators, which now includes blog posts, Twitter streams, and other forms of media, as the "World News Connection." These translations, which collectively represent the most international newspaper known to humankind, are available for an annual subscription fee of $300 plus $4 per article retrieved.[16]

Unsurprisingly, World News Connection is a tough sell, because of the expense and because most readers—even passionate Azerbaijan watchers—don't want every story produced by Baku's newspapers. Translators like Roland Soong—the man who translated the *Qilu Evening News* article for his readers— are valuable not just because they produce text that's easy to read but also because they act as filters and select stories that are likely to be interesting to a broader audience.

Roland Soong and the Future of Translation

A professional media researcher who studies the size and demographics of mass media audiences around the world, Soong relocated from New York to Hong Kong in 2003 to spend more time with his elderly mother. Thrust into a Chinese-language media world, Soong felt compelled to catch up, and he quickly discovered

> that Chinese-language and English-language readers were getting different kinds of news. Many things of interest to the Chinese were filtered out or simplified for various reasons

(such as cultural barriers, target audience needs, space, political bias, etc.). So I began to look for the most interesting instances in Chinese and translate them to English so that English-only readers can have a better understanding of the issues and backgrounds.[17]

Soong posts these translations to EastSouthWestNorth, a website whose stark design almost disguises the wealth of content it contains. ESWN's homepage includes headlines in three columns: Global, Greater China (in English), Greater China (in Chinese). The left column follows the work of columnists and scholars, commenting on China and on broader world issues, while the right lists stories in Chinese publications that are getting attention in China. The middle is where Soong's hard work is most visible. Several articles a day, sometimes totaling thousands of words, are selected from Chinese publications and translated into English by Soong, who spends anywhere from thirty minutes to six hours a day translating.

The motives for translating a specific story vary, but the general operating principle is that these stories are important to Chinese audiences but invisible to the rest of the world.

It may be a story that has almost all of China involved, but there is scarcely any reaction outside China. The reasons may be cultural, political (usurps western narrative) or substantive (too complicated), but I will translate it if I think it tells people about what is important in China. . . . It may be a follow-up on a story that was reported in western media at first, but later evolved into something different which was not followed up. With the Internet today, many stories require investigative efforts to confirm, but people don't like to be told that they had been initially misled.

What becomes clear when discussing translation with Soong is that the model of a remote China, isolated from the rest of the world behind "the Great Firewall," is insultingly simplistic. Yes, Chinese censors are quite effective at preventing some stories, like accounts of political upheaval in Tunisia and Egypt in early 2011, from gaining wide exposure inside China. But censors' efforts are far more often focused on preventing stories about corruption in one corner of the vast nation from being reported in other cities, for fear the stories might inspire public demonstrations. By translating those stories into English, Soong invites international reporters to explain tensions around power and control in China to their audiences . . . and sometimes to Chinese readers as well.

Soong was one of the few sources of information in English about a wave of protests that began in Taishi Village in Guangzhou in August 2005. Attempts to oust the corrupt village committee director Chen Jinsheng led to hunger strikes, sit-ins, the arrest and savage beating of the activist Lu Banglie, and the deployment of 1,000 riot troops to subdue a village of 2,075 peasants. Chinese media covered the story extensively through September, and Soong translated much of the coverage. By early October, the Taishi story was getting heavy coverage in Asian papers like the *South China Morning Post*, but it hadn't appeared in major American papers. That changed when the *Guardian* reporter Benjamin Joffe-Walt accompanied Lu to Taishi and was detained by local authorities; Joffe-Walt offered a sensationalistic account of a beating administered to Lu, and the *Guardian* was forced to retract and correct his original report. Joffe-Walt's story and Lu's detention gained attention through the controversy and brought reporting on two months of protests in Taishi to US and UK audiences.[18]

While countless American commentators, most notably Secretary of State Hillary Clinton, have criticized China's firewall and decried Chinese censorship, far fewer have pointed out that there's lots of potentially important, uncensored Chinese news that never reaches an English-speaking audience. China's censored press provided a great deal of information about Taishi, at least in early stages of the protests. Soong translated an opinion piece from the

People's Daily, the official newspaper of the Chinese Communist Party, supporting the protests, telling readers, "This is akin to official blessing by the central government." The Taishi story, in an optimistic first act of successful village defiance and a sad second act of government crackdown, has been an interesting and revealing instance of governmental change in China. That Taishi isn't familiar to non-Chinese readers is a function of shortcomings in Western media, not primarily of Chinese censorship.

Soong's quest to reveal what's important to China to an international audience has gained fellow travelers. "Blogs such as chinaSMACK and ChinaHush are covering many of the social stories that I used to do plenty of," Soong notes, which leaves him free to focus on topics that fascinate him: media reporting accuracy, ethics, and manipulation. His site continues to publish multiple stories and thousands of words in translation a day.

Others have joined Soong in his quest to make Chinese-language media accessible to global audiences. Tea Leaf Nation, an e-magazine produced by three friends who met at Harvard—two are Chinese, one is an American who learned Chinese as a Peace Corps volunteer—translates political stories from social media for English readers. Ellen Lee and Casey Lau produce "Weibo Today," a weekly YouTube video segment that reports on trends in China's microblogging services, or "weibos." But the people behind these projects are vastly outnumbered by Chinese translators working to make the English-language Internet accessible to an audience of over 400 million Chinese-speaking Internet users.

The Internet entrepreneur Zhang Lei began translating from English to Chinese for the most personal of reasons: his father's death from lymphoma in 1996, the year Zhang came to the United States as a student. "Since then I have been keeping watch on materials about this disease in both Chinese and English on and off," he has said. "What struck me the most was that in English literature lymphoma had been considered a curable disease, however that critical piece of information was not available for Chinese patients. This motivated me to discuss possible solutions to this problem with my friends."

Inspired by projects like Wikipedia, Zhang and two friends began a project to allow people to work collaboratively on translations. Yeeyan, their group translation site, was born in 2006. It began to grow in earnest amid rising tensions between the United States and China before the 2008 Olympics. Watching US media coverage oscillate between the construction of stadiums, questions about China's human rights record, and clashes between Uighur protesters and soldiers in Urumqi, in western China, Zhang saw clear evidence of the ways in which Chinese and American audiences fail to understand each other.

"I didn't know what I could do," he said in a presentation at the 2009 Chinese Internet Research Conference at the University of Pennsylvania. "But I knew we could translate." Yeeyan has over 210,000 registered volunteer translators who work together to translate key English-language media into Chinese. Collectively, they average a thousand translated stories per week. The contents vary, but on an average day Yeeyan.org features stories from major newspapers like the *Guardian* or the *New York Times*, from weekly news magazines like *Time* or *Newsweek* (an unrelated team, the Ecoteam, works on translating the *Economist* into Chinese each week)[19] and prominent websites like *ReadWriteWeb*. They've taken on the translation of books as well, such as the US Federal Emergency Management Agency's *Earthquake Search & Rescue Manual* and *Earthquake Safety Manual* in the aftermath of the 2008 Sichuan earthquake. And close to Zhang's heart, the group has translated a book called *Getting Started with Lymphoma*, which has been downloaded by over 100,000 Chinese readers.

Yeeyan is likely to face complex copyright issues in the long run, since some authors translated by Yeeyan may not want their content translated into Chinese, especially if Yeeyan begins syndicating content to Chinese newspapers and websites. But other publishers have embraced the project enthusiastically. The *Guardian* began pointing to Yeeyan's translations as its official Chinese version in 2009, though it was forced to end the partnership shortly afterward.

If copyright hasn't yet proven a major stumbling block for Yeeyan, censorship has. Unlike Soong, who translates into English Chinese-language content that has already been published in China, some of Yeeyan's English-language sources are regularly blocked in China. Government officials shut the site down in December 2009 when they grew concerned that the content translators were posting violated local content guidelines. These guidelines change quickly and are often ambiguous, but they are a fact of life for Chinese media companies. After a difficult internal debate, Zhang and his team decided to bring Yeeyan into compliance with local self-censorship requirements. A team now reviews translations and stops the publication of content likely to cause the project to be blocked. "We communicate to our translators on individual basis when sensitive translations can not be published and will only be stored as draft for translators' own record. This treatment is sadly the de facto standard for UGC [user-generated content] sites operating within China, therefore is accepted by community members," explains Zhang.

Inspiring as Yeeyan is, the site also poses a discomfiting question: Where's the English-language equivalent? The 210,000 volunteers believe it's important for Chinese audiences to understand what English-language media are saying, and those volunteers donate time to bridge barriers of language. Thousands of others are involved in less-weighty types of translation, subtitling English-language movies and television shows on sites like yyets .com. I find it hard to believe that the Chinese-language Internet, where roughly half of the over 400 million users create content via blogs or microblogging services, produces so little content that Roland Soong and a few dozen others can translate all that's potentially interesting to English-speaking audiences.

Of course, Yeeyan has an advantage over a parallel English-language project, because many university degrees in China require proficiency in English, creating a ready population of potential translators. But we've not seen a project emerge in the United States at the scale of Yeeyan to translate Spanish-language content, for instance, even though many US high school students learn the language and though a significant percentage of the US population speaks Spanish as a first language and creates online content in that tongue.

What amazes many people about Yeeyan—the willingness of translators to work on a project like Yeeyan without direct financial compensation—is well explained by scholars of open-source software and of Wikipedia. While it's possible for very experienced translators to make a living translating online, and while a larger set makes a few cents at a time translating through online labor markets like Amazon's Mechanical Turk, for Yeeyan's translators, it's closer to being an avid hobby than a job. Zhang explains that a number of other motivating factors come into play. Translators are looking for experience, which they might leverage into well-paying jobs. But they're also motivated by recognition from the community, a sense of accomplishment that comes from improving as a translator and from enjoyment of the material they're translating. The same motivations that make community projects like open-source software and Wikipedia possible seem to work with human translation. It's a gift culture, where status comes from providing the best gifts, the most helpful translations. And giving conveys status. In his seminal *Wealth of Networks*, Yochai Benkler recognizes this as "agonistic giving—that is, giving intended to show that the person giving is greater than or more important than others, who gave less."[20]

Other communities that have succeeded with online translation have embraced aspects of this model. The invitation-only TED—Technology, Education, Design—conference began reaching beyond the few thousand attendees who see talks delivered live (in English) in Monterey, California, when TED's media producer June Cohen started publishing videos of the talks on the Internet in 2006.[21] After three years of publishing videos, June realized that many more people would be interested in the lectures if they could watch them with subtitles in their native language. She began paying a transcription firm to produce high-quality English transcripts of the talks and hired professional translators to produce subtitles in Tagalog or Turkish.

Inspired in part by Global Voices' success in using volunteers to translate our online reporting, June began an experiment: she invited volunteers to translate some talks, and hired professionals to work on others, in order to set a high bar for quality. "As it turns out, the volunteers have consistently produced translations that are as good or better than those we paid for," said June. "We were amazed." TED translators are not monetarily compensated for their work, but they are widely celebrated, given near-equal billing on the website to the speakers themselves, and the most prolific and successful translators are invited to attend the conferences in person. What makes the project work, June believes, is the combination of community recognition of the importance of the effort and the fact that translators choose what material they work on. "Translating a talk you're passionate about is fun, while translating one that bores you is work." As a result, the volunteer translation model works best when the goal isn't to translate everything but to prioritize the most compelling material.

The power and reach of volunteer translation is surprising. Al Gore's 2008 TED talk on global warming, over an hour long, has been translated into thirty-six languages and watched by 1.5 million viewers. Two years into the project, TED translators produced 18,000 translations in eighty-one languages. The average talk is translated into twenty-four languages within a few weeks. While only about 10 percent of TED.com viewers watch talks with non-English subtitles, that's more than a million viewers a month. And TED partners with Youku, the Chinese YouTube competitor, to show TED talks with Chinese subtitles to millions more viewers.

Translation via the volunteer models is powerful, but it's not quick. It may allow Arabic speakers to understand an English-language lecture, but they must wait days or weeks until an Arabic translator takes on the task of localizing the piece. And the translation of the talk doesn't help them participate in the online discussions that erupt when a new talk is posted. What we really want are translations that are as nuanced and accurate as those produced by TED or Global Voices translators and as fast as Google Translate.

Ed Bice's Meedan.net project is an online space where Arabic and English speakers can interact on common linguistic ground, thanks to both machine and human translation. The word "Meedan" means "public square" in Arabic, and the project attempts to create an online public space for conversations between English and Arabic speakers, a small online community focused on discussing current events in the Middle East in Arabic and English. Articles posted to the site from online news sources are automatically translated between Arabic and English by means of machine translation. Comments on stories can be posted in English or Arabic, and they are automatically translated after posting. These machine translations are considered the first step within the Meedan community; volunteers look for new stories or comments and "clean up," or sometimes thoroughly revise, the machine translations that have been posted. Machine translation allows a conversation to unfold in real time, involving speakers of two languages. Human translation makes that conversation more readable and creates a permanent record that turns the conversation into an online resource.

If Bice's hopes for increasing English/Arabic dialogue through translation are ambitious, they pale in comparison with Luis von Ahn's plans for Duolingo. Von Ahn, a professor at Carnegie Mellon University, is an expert in the emerging field of "human computation." Human computation uses the skills of thousands of human beings, working in parallel, to solve problems that are difficult for computers to solve. Von Ahn is best known for the "reCAPTCHA," which you've likely seen on the comment form of websites. The reCAPTCHA asks you to transcribe two words to demonstrate that you are a human being and not a computer program. In the process, you're helping transcribe scanned book pages, one word at a time. reCAPTCHAs were transcribing the equivalent of 160 books per day in 2008[22] and are now correcting errors in the Google Books project, Google's vast effort to scan the contents of major university libraries.

If humans can decipher ambiguous scanned words and transcribe books, why can't they translate documents? Von Ahn asked his graduate student Severin Hacker, "How can we get 100 million people translating the Web into every major language, for free?"[23] The answer they came up with involves encouraging millions of people to learn a second language. Join Duolingo, and you'll be invited to learn Spanish, French, or German. The sentences you translate at first are simple and formulaic, but as you improve, you'll be invited to translate sentences from live web pages.

Should you trust someone who's just learning Spanish to translate a web page for you? Von Ahn's algorithms help merge the judgments of dozens of inexperienced translators into a translation he claims rivals the quality of a professional translator's. Scale is on his side, because 30 million people solve reCAPTCHAs a day. If only a small percentage of those users decide to learn a new language, von Ahn believes, he'll be able to translate the English Wikipedia into Spanish in less than a week.[24]

Digital Extinction?

While Yeeyan and TED prove that volunteers can produce high-quality translation of newspaper stories and academic lectures, and Meedan suggests that a combination of machine and human translation could enable real-time communication across languages, the really exciting possibility comes in bringing these methods together. For machine translation to work, programmers need large corpora of material translated between a pair of languages. While the amount of text translated by Global Voices or TED is currently a small fraction of the text necessary to build a statistical machine translation system, a partnership between translation communities and machine translation experts might generate corpora where few other options exist. The four thousand translations produced by Global Voices Malagasy, totaling 300,000 words, amount to only 1.2 percent of the size of the Europarl corpus (one of the key sources for parallel corpora, derived from European parliamentary proceedings)[25] and are likely too small for an accurate machine translation system. On the other hand, it's probably the largest available corpus that translates between English and Malagasy.

Google's ambitions to index and make available all the world's knowledge means it has to take seriously the existence of any corpora for African languages. For the vast search engine to keep growing internationally, it must provide services to hundreds of milions of people for whom English, French, and Portuguese are second languages. According to Denis Gikunda, who leads African-language initiatives for the company, Google plans to offer translation services, interfaces, and content in over one hundred African languages that have at least one million speakers, including Meru, his native tongue, which is spoken in the area near Mount Kenya.[26] For now, Google is focusing on bigger languages—Swahili, Amharic, Wolof, Hausa, Afrikaans, Zulu, Setswana, and Somali, all of which have at least ten million speakers.

In order for Google or others to translate Malagasy, they need more than a set of pages translated between English or French and Malagasy; they require piles of data to build a Malagasy "language model." In other words, for Malagasy to be translatable through statistical machine translation, a great deal written in Malagasy must be available online or easily digitized. That presents a problem. Consider the Malagasy Wikipedia, which contains roughly 25,000 articles. That makes it the seventy-fifth-largest Wikipedia in the world, and the second-largest in an African language. Many of the potential contributors to the project are well-educated people from Madagascar who also speak French fluently. The French Wikipedia has fifty times as many articles and a vastly larger audience. A Wikipedian looking to have her contribution read and appreciated is likely to contribute in French.

Lova Rakotomalala, a contributor to the Malagasy Wikipedia, explains the catch-22: "My hunch is that people are not using the (smaller language versions of) Wikipedia because of a vicious cycle. People don't want to create the content because no one is reading, and no one is reading because there is no content." Like the Jordanian bloggers who wrote in English to reach a global audience, Malagasy speakers have a strong incentive to write in French. And unless they write in their native language, there may not be a tipping point, as there was in the Arab blogosphere.

This would be more discouraging if Rakotomalala weren't deeply engaged in expanding the amount of Malagasy content available online, both through Wikipedia and through Global Voices, where he cofounded our Malagasy-language edition. But his comment helps elucidate the complicated issues that surround hopes for a polyglot Internet. If Malagasy speakers post more content online, more Malagasy speakers are likely to create content in their native language. With more content—and especially more content in translation—online, Google and others may be able to build machine translation systems, which in turn means that content available only in Malagasy can be read by people who don't know the language.

If Malagasy speakers decide, instead, to create content in

French, looking for a larger audience, they may suffer another problem. Projects like the English- and French-language Wikipedias are reaching "maturity"; the sites contain so many articles that experienced editors now reject at least as many new articles as they accept. Articles on important aspects of Madagascar's geography, fauna, and culture may be enormously significant to people in that country, but might not meet Wikipedia's "notability" threshold for inclusion in the French Wikipedia. In a Malagasy Wikipedia, local knowledge is an obvious candidate for inclusion; in a larger, more global Wikipedia, the same information might not merit an article.

The existence or nonexistence of a single article in Wikipedia may not represent a cultural crisis. But the extinction of languages merits our attention. The anthropologist Wade Davis notes that half of the world's six thousand languages are no longer being taught to schoolchildren. Without another generation of native speakers, most will die out.[27] Those concerned about language extinction worry that culturally dominant neighbors will force out smaller languages. The five million speakers of Mayan often also speak Spanish, a global language. It's not hard to imagine those speakers deciding it's to their economic advantage to speak Spanish for the most part and let Mayan slowly disappear.

The cases we're considering here outline the force the digital world can exert on language disappearance. If speakers don't have an incentive to create content in a language, we won't have enough content online to build translation models. The bits of Malagasy or Mayan content online may remain linguistically "locked up," available only to native speakers and invisible to everyone else. We may be facing a wave of digital language extinction, where some languages have a large enough online presence to maintain a community and develop a machine translation system, while others fall beneath that threshold and never make a significant mark online.

Making Translation Transparent

The ability to translate a language, via automated systems or volunteer translators, doesn't guarantee that we'll ever encounter those translations. Roland Soong's translation of the *Qilu Evening News* story about Lanxiang Vocational was available online, but the journalists writing about Chinese hackers didn't find it. Search is such an important discovery mechanism that for many of us, information not easily accessible via a search engine doesn't exist. Crossing the barrier of language requires more than making translation merely possible. It involves making language transparent.

At moments of crisis, we're often reminded how powerful language barriers can be. As Tunisia, Egypt, and then much of North Africa and the Middle East exploded into popular protest in early 2011, many fascinated readers turned to Twitter for real-time reporting and commentary. Much of what was most interesting on Twitter was written in Arabic, not in English. Some extraordinary reporters, like Dima Khatib, Al Jazeera's Latin America bureau chief, acted as real-time translators, posting in Arabic, English, and Spanish to reach a broad swath of users.

Andy Carvin, NPR's strategist for social media, put all other work aside for the first months of 2011 and dedicated himself to covering these struggles via online media. His Twitter feed, followed by over 25,000 readers around the world, frequently included pleas for help in translating a slogan shouted in Tahrir Square or a tweet by a Tunisian dissident. Since such a wide audience followed his aggregation, translations often appeared seconds later, and Carvin immediately reposted the information. Danny O'Brien, an advocate for online free speech with the Committee to Protect Journalists, automated the process by writing a simple tool—a web browser extension—that adds a "translate" button to Twitter beside each tweet, allowing an interested reader to quickly read a machine translation of an otherwise unreadable post.

Carvin's and O'Brien's methods work well when we're motivated to seek out content in another language. But we're still more likely to decide to follow a Twitter friend who speaks in a language we understand. Until language becomes entirely transparent, it will shape whom we choose to listen to and whom we ignore.

Google is taking steps to try to make translation more transparent across its products. When you load a page in Google's Chrome web browser, the program tries to detect what language it's written in and, if the language isn't your default, offers a machine translation of the content. You can disable this feature, accept translations as they're offered, or tell Chrome always to translate content in a particular language into your native tongue. My installation of Chrome now renders pages in Chinese, Japanese, and Arabic into English for me by default, and I've discovered that I no longer instinctively reach for the back button in my browser when I stumble off the comfortable path of English-language pages. The translations offered can be hard to read, but at minimum I have a sense of what topics they're covering and whether I might beg a multilingual friend for a more readable translation. Google's Gmail service now operates in a similar way, offering to seamlessly translate email you receive written in languages you don't read.

But Google faces a larger challenge. For language to recede as a major barrier online, translation needs to move from the browser into the search engine. When we look for information through most search engines, the language we use to build a query limits the results we get. Search Google in the United States for "apple" and you won't get the same results as you would get by searching for the Spanish equivalent, "*manzana*," on Google.mx. This makes sense, of course—many of the people searching in the

United States would prefer English-language results. But this limitation can constrain what information is available.

Ivan Sigal, Global Voices' executive director, is a serious amateur cyclist. When he bought a secondhand, handmade bicycle frame made by an obscure, defunct German manufacturer called Technobull, he immediately wanted to learn more about his new ride and other folks who rode the same brand. Searching for information on Google.com, he found virtually nothing, a few dozen pages in English referencing the brand as elite and expensive and one page of Flickr images. So he turned to Google.de and discovered thousands of pages, including an active online forum of riders who venerate Technobull cycles. Ivan speaks a little German, and some of the riders were willing to answer his questions and help him out. The information Ivan needed was in German, not English, and Google.com wasn't able to help him find what he needed.

Yet, Google's director of product management, Anjali Joshi, wants to make sure that language isn't an insurmountable barrier to sharing knowledge. "A person in Korea, or any part of the world, should have access to all information on the web in their language, rendered perfectly, in a way that's readable, understandable, findable." This goes beyond a search for apples that looks for results in English, Spanish, and Korean: "Eventually we want people to be able to converse with each other, to move seamlessly between languages in spoken and written language."

Despite Google's dramatic progress in translation, this remains a long path. (Google can translate between English and sixty other languages. Translations from Icelandic to Yiddish, for instance, go through English as a "bridge language.") "There are three keys to getting there," Joshi tells me. "We need excellent machine translation first. Then we need perfect search results across all languages." In other words, we have to have search algorithms that can decide whether a translation of a Spanish page on *manzanas* or an English page on apples was the better search result.

Joshi's colleagues, sitting with us in a Mountain View conference room, look slightly nervous about the challenges of achieving her vision as she leans back in her chair and offers the third
step. "Once you can search in every language, and you have perfect translation, you have the best content for everyone on the
web. That would be Nirvana."

I think Joshi is partly right. Even if we have perfect, multilingual search, we face another challenge: understanding the implications and importance of what a person is saying. To discover
information from different parts of the world, we need more
than excellent translation: we must understand the context of
the words we encounter. The path toward Nirvana is a long one,
and along the way we need guides who can put our discoveries in
context.

Chapter 6

TAKEN IN CONTEXT

THE EARLY 1980S WEREN'T ESPECIALLY KIND TO PAUL SIMON. He ushered in the second decade of his post–Simon & Garfunkel life with *One Trick Pony*, a forgettable companion album to a forgettable film starring his former musical partner, Art Garfunkel. When a 1981 reunion concert with Garfunkel brought 500,000 people to New York's Central Park, and sold over two million albums in the United States, the two began touring together. But "creative differences" brought the arrangement to a premature end, and a planned Simon & Garfunkel album became a Simon solo release, *Hearts and Bones*, that was the lowest-charting of his career. With the breakup of his marriage to the actress Carrie Fisher, "I had a personal blow, a career setback and the combination of the two put me into a tailspin," Simon told his biographer Marc Eliot.[1]

During this dark period, Simon was mentoring a young Norwegian songwriter, Heidi Berg. Berg gave Simon a cassette of mbaqanga music featuring musicians from Soweto, then the most notorious blacks-only township in apartheid South Africa. While the identity of the album Simon heard is uncertain,[2] it likely featured the Boyoyo Boys, a popular Sowetan band, and listening to the cassette in his car, Simon began writing new mel-

ody lines and lyrics on top of the sax, guitar, bass, and drums of
their existing tracks.

"What I was consciously frustrated with was the system of sitting and writing a song and then going into the studio and trying
to make a record of that song. And if I couldn't find the right
musicians or I couldn't find the right way of making those tracks,
then I had a good song and a kind of mediocre record," Simon
told *Billboard* magazine's Timothy White. "I set out to make
really good tracks, and then I thought, 'I have enough songwrit-
ing technique that I can reverse this process and write the song
after the tracks are made.'"[3]

In the hopes of working this new way, Simon appealed to his
record company, Warner Bros., to set up a recording session
with the Boyoyo Boys. In 1985, that was far from an easy task.
Since 1961, the British Musicians Union had maintained a cultural boycott of South Africa, managed by the UN Center against
Apartheid. The boycott was designed to prevent musicians
from performing at South African venues like Sun City, a hotel
and casino located in the nominally independent bantustan of
Bophuthatswana, an easy drive from Johannesburg and Pretoria.
But the boycott covered all aspects of collaborations with South
African musicians, and Simon was warned that he might face
censure for working in South Africa.

When Simon turned to Warner Bros. for help, the company called Hilton Rosenthal. Then managing an independent
record label in South Africa, Rosenthal had in the past worked
with Johnny Clegg and Sipho Mchunu, the two musicians who
became the heart of Juluku, a racially integrated band that elec-
trified traditional Zulu music and brought it to a global audience.
Rosenthal's label had partnered with Warner Bros. to distribute
Juluku's records in the United States, so Warner executives knew
he could help Simon navigate a relationship with South African
musicians.

As a white South African who'd recorded a highly political,

racially integrated band in apartheid Johannesburg, Rosenthal was aware of some of the difficulties Simon might face in recording with Sowetan musicians. He assured Simon that they would find a way to work together and sent him a pile of twenty South African records, both mbaqanga acts and choral groups including Ladysmith Black Mambazo. Then he arranged a meeting with his friend and producer Koloi Lebona, who set up a meeting with the black musicians' union, to discuss whether members should record with Simon.

The musicians had reason to be skeptical of such a collaboration. They had previously crossed paths with one of popular music's great appropriators, Malcolm McLaren. McLaren is best known as the Svengali behind the Sex Pistols, assembling the seminal punk band at his London clothing boutique. The controversial, explosive, brief, and ultimately tragic career of the Sex Pistols launched McLaren as a musical innovator and provocateur.

For his next act, McLaren didn't bother to build a band. *Duck Rock*, released in 1983, is a complex and compelling pastiche of influences from around the globe: American folk, early hip-hop, Afro-Caribbean, and lots and lots of mbaqanga music. "Double Dutch," an ode to African American jump rope culture, is built around an instrumental track, "Puleng," by the Boyoyo Boys. McLaren didn't credit the Boyoyo Boys for the track, claiming he'd authored it with the Yes bass player Trevor Horn.[4] The album borrowed heavily from other South African acts, including Mahlathini and the Mahotella Queens, who also worked unpaid and uncredited.

When Simon approached Rosenthal about recording with the Boyoyo Boys, the band was in the early stages of a lawsuit attempting to get royalties from McLaren.[5] But Rosenthal and Lebona encouraged the collaboration, and a majority of the black musicians' union agreed to invite Simon to South Africa to record. They worried that the UN cultural boycott was preventing mbaqanga music from taking its place on the global stage, as

Jamaican reggae had done. Realizing that Simon's stature could bring a great deal of attention to the local musical scene, they voted to work with him.

The sessions that Rosenthal and Lebona organized led to *Graceland*, one of the most celebrated albums of the 1980s. It won Grammy awards in 1986 and 1987, topped many critics' charts and regularly features on "top 100 albums of all time" lists. It also made a great deal of money for Simon and the musicians he worked with, selling over sixteen million copies. South African songwriters share credits and royalties with Simon on half of the album's tracks, and Simon paid session musicians three times the US pay scale for studio musicians. Many involved with the project, including Ladysmith Black Mambazo, drummer Isaac Mtshali, and guitarist Ray Phiri went on to successful international music careers.

At its best, *Graceland* sounds as if Simon is encountering forces too large for him to understand or control. He's riding on top of them, offering free-form reflections on a world that's vastly more complicated and colorful than the narrow places he and Art Garfunkel explored in their close harmonies. The days of miracle and wonder Simon conjures up in "The Boy in the Bubble" are an excellent metaphor for anyone confronting our strange, connected world.

Collaborations like *Graceland* don't happen without the participation of two important types of people: xenophiles and bridge figures. Xenophiles, lovers of the unfamiliar, are people who find inspiration and creative energy in the vast diversity of the world. They move beyond an initial fascination with a cultural artifact to make lasting and meaningful connections with the people who produced the artifact. Xenophiles aren't just samplers or bricoleurs who put scraps to new use; they take seriously both forks of Kwame Appiah's definition of cosmopolitans: they recognize the value of other cultures, and they honor obligations to people outside their own tribe, particularly the people they are influenced and shaped by. Simon distinguishes himself from McLaren by engaging with South African musicians as people and by becoming an advocate and promoter of their music.[6]

Unlike xenophiles, outsiders who seek inspiration from other cultures, bridge figures straddle the borders between cultures, figuratively keeping one foot in each world. Hilton Rosenthal was able to broker a working relationship between a white American songwriter and dozens of black South African musicians during some of the most violent and tense moments of the struggle against apartheid. As a bridge, Rosenthal was an interpreter between cultures and an individual both groups could trust and identify with, an internationally recognized record producer who was also a relentless promoter of South Africa's cultural richness. Rosenthal, in turn, credits Koloi Lebona with building the key bridges between black musicians and the South African recording community.

Bridge Figures

The Chinese activist and journalist Xiao Qiang and I started using the term "bridging" to describe the work bloggers were doing in translating and contextualizing ideas from one culture into another. Shortly afterward, the Iranian blogger Hossein Derakhshan gave a memorable talk at the Berkman Center as part of the Global Voices inaugural meeting. Hossein explained that, in 2004, blogs in Iran acted as windows, bridges, and cafés, offering opportunities to catch a glimpse of another life, to make a connection to another person, or to convene and converse in a public space. I've been using the term "bridgeblogger" ever since for people building connections between different cultures by means of online media, and "bridge figures" to describe people engaged in the larger process of cultural translation, brokering connections and building understanding between people from different nations.

To understand what's going on in another part of the world often requires a guide. The best guides have a deep understanding of both the culture they're encountering and the culture they're rooted in. This understanding usually comes from living for long periods in close contact with different cultures. Sometimes this is a function of physical relocation—an African student who pursues higher education in Europe, an American Peace Corps volunteer who settles into life in Niger semipermanently. It can also be a function of the job you do. A professional tour guide who spends her days leading travelers through Dogon country may end up knowing more about the peculiarities of American and Australian culture than a Malian who lives in New York City or Sydney but interacts primarily with fellow immigrants.

My friend Erik Hersman is an American, a former Marine, who lives and works in Nairobi, Kenya. The child of American Bible translators, Hersman grew up in southern Sudan and in the Rift Valley of Kenya. After school and military service, Erik ran a technology consultancy in Orlando, Florida, making regular trips to East Africa to document technological innovation on the blog *Afrigadget*. He then moved to Nairobi to lead the *iHub, a technology incubator in central Nairobi designed to nurture Internet-based start-ups.

Erik is able to do things most Americans aren't able to do. He can wander around Gikomba in Nairobi and talk to local metalworkers in Swahili for a blog post about African hacking, because he's a Kenyan. And he can help Kenyan geeks develop a business plan to pitch a software venture to international investors because he's an American geek. Lots of people have one of these skill sets, but bridge figures are lucky enough to have both.

The sociologist Dr. Ruth Hill Useem uses the term "third culture kid" to describe individuals like Erik who were raised both in the home culture of their parents and in the culture of the places where they grew up. Useem argues that kids raised in this way end up developing a third culture by combining elements of their "birth" culture and the local culture they encounter. Children who go through this process—the kids of military personnel, missionaries, diplomats, and corporate executives—often have more in common with each other than with other kids from their birth culture. Researchers working in the same vein as Useem's have found evidence that some third culture kids are often well adapted to live and thrive in a globalized world. Frequently they're multilingual as well as multicultural, and are very good at living and working with people from different backgrounds. As a downside, some third culture kids report feeling that they're not really at home anywhere, in either their parents' culture or the culture they were raised in.

While Useem's research focuses primarily on North Americans and Europeans growing up in other parts of the world, international patterns of education and migration are giving people from many nations the opportunity to become bridge figures. Hundreds of the individuals who write or translate for Global Voices are citizens of developing nations who've lived or worked in wealthier nations, learned new languages, and absorbed new cultures as students, migrants, or guest workers.

Merely being bicultural isn't sufficient to qualify you as a bridge figure. Motivation matters as well. Bridge figures care passionately about one of their cultures and want to celebrate it to a wide audience. One of the profound surprises for me in working on Global Voices has been discovering that many of our community members are motivated not by a sense of postnationalist, hand-holding "Kumbaya"-singing, small-world globalism but by a form of nationalism. Behind their work on Global Voices often lies a passion for explaining their home cultures to the people they're now living and working with. As with Erik's celebration of Kenyan engineering creativity, and Rosenthal's passion for the complexity and beauty of South African music, the best bridge figures are not just interpreters but also advocates for the creative richness of other cultures.

The Taliban, McDonald's, and Curried Goat

What happens when people encounter another culture for the first time? Will we find a bridge figure to help us navigate these encounters? How often do we embrace the unfamiliar as xenophiles, and how often do we recoil and "hunker down," as Robert Putnam observes?

It's a question as old as the *Odyssey*, where Odysseus's encounters with people of other lands remind readers that his name, in Greek, means "he who causes pain or makes others angry."[7] For all the kindly Phaeacians who sail Odysseus back to Ithaca, there are Cyclopes who eat men and destroy ships. When we encounter new cultures, should we expect cooperation or conflict?

The political scientists Pippa Norris and Ronald Inglehart consider this ancient question through the lens of media. In their book *Cosmopolitan Communications*, they look at what happens when people encounter different cultures through television, film, the Internet, and other media. Their exploration starts by examining the introduction of television to the small, isolated Buddhist nation of Bhutan in 1999. Prior to 1999, television had been illegal in Bhutan, though a small number of people had televisions and rented Hindi-language videocassettes to watch at home. In June 1999, King Jigme Singye Wangchuck allowed Bhutanese to begin watching television and to connect to the Internet. Two Bhutanese businessmen soon formed Sigma Cable, which by May 2002 offered forty-five Indian and American channels to about four thousand households.[8]

Almost immediately after the introduction of television, Bhutanese journalists started reporting on an apparent crime wave, including drug offenses, fraud, and murder. Bhutanese schoolchildren began watching professional wrestling and practicing body slams on fellow students in the schoolyard. The situation escalated into a moral panic, as citizens and journalists speculated that television morality would overwhelm Bhutanese values and traditions.

Bhutanese authorities had hoped that a local public broadcaster, charged with producing educational content about Bhutanese traditions, would help temper the influence of foreign media. But the broadcaster was slow to produce programming, and the Hindi soap operas and British news programs offered via cable television were far more popular. By 2006, the government had created a new ministry to regulate media, which promptly banned sports and fashion channels as well as MTV on the grounds that they had "no suffering alleviation value."[9] Worried that television was teaching young Bhutanese to stay at home and watch soap operas, the nation's health and education minister embarked on a fifteen-day, 560-kilometer trek to warn his citizens against indolence: "We used to think nothing of walking three days to see our in-laws. Now we can't even be bothered to walk to the end of Norzin Lam high street."[10]

Television's apparent transformation of Shangri-la into a land of violent, criminal couch potatoes expresses one set of fears associated with cross-cultural encounter. Western media are so powerful and insidious, this argument goes, that a fragile culture like Bhutan's can't possibly hope to compete. Faced with *American Idol*, Coca-Cola, and McDonald's, Bhutan's culture will inevitably capitulate to the dominant, Western culture unless governments aggressively intervene.

Norris and Inglehart argue that there are at least three other possible outcomes to these types of encounters: resistance, synthesis, and disengagement. We might see one culture violently reject another, which they term "the Taliban effect." The banning of Western music and movies in Taliban-controlled parts of Afghanistan and the violent opposition to secular education in northern Nigeria by Boko Haram are both examples of the ways encountering another culture might lead to polarization instead of extinction at the hands of a dominant culture. So too can dominant cultures polarize in the face of perceived invasion or threat: when the city of Nashville, Tennessee, tried to ban the use of languages other than English in city buildings, it signaled a retreat from tolerance in the face of the perceived threat of immigration.[11]

Happier possibilities exist. We can imagine "a blending of diverse cultural repertoires through a two-way flow of global and local information generating cross-border fertilization, mixing indigenous customs with imported products."[12] Consider curry, where encounter between the food of the Indian subcontinent and the rest of the world has led to syncretic cuisine like Japanese kare-pan (curry-stuffed bread), Trinidadian curried goat, or that paragon of British cuisine, the curry jacket potato. Cultural encounter can lead to creative fusions that honor both cultures while creating something unexpected and new.

We could also encounter another culture, shrug our collective shoulders, and conclude, "That's not for us." Norris and Inglehart call this "the firewall theory," and suggest that deeply rooted cultural attitudes and values are quite robust when confronted with other cultures through the flows of media and communication. These values act as a "firewall," allowing some influences to pass through and others to be filtered out. The researchers find ample evidence that cultural values—as measured by instruments like the World Values Survey—are quite slow to change, even when countries are well connected through media technologies. South Africa, for instance, has become much more connected to global media and economics since the fall of apartheid, but the World Values Survey finds evidence for the survival of conservative social values during this period of sharp change.[13]

This finding is good news for anyone concerned about the youth of Bhutan. It's also consistent with the effects of homophily on social and professional media. Being connected to global flows of information doesn't guarantee that we'll feel their influence over and above the influences of homegrown media, or the preferences of friends and family. But it presents a challenge to those who believe that cultural encounter can lead to outcomes as banal as revitalized pop careers and improvements in snack food, or as significant as novel solutions to global problems like climate change. Creative fusion may happen by accident, but it's far from guaranteed. If we want the benefits that come from sharing ideas across borders, we need to work to make it happen.

Weak Ties or Bridge Ties?

Who's most likely to help you find a new job—a close friend you talk to every week, or an acquaintance you see a few times a year? The close friend has more motivation to help with your job search, but he probably knows many of the same people you do. The acquaintance has connections to different social networks and is likely to know of opportunities you haven't already encountered. In fact, many important contacts come through people whom job seekers barely know or have fallen out of touch with—old college friends, former colleagues. That's the conclusion the sociologist Mark Granovetter reaches in his widely cited paper "The Strength of Weak Ties." In his words, "It is remarkable that people receive crucial information from individuals whose very existence they have forgotten."

Granovetter's finding has been so widely popularized that it's become standard job-seeking advice. The popular social-networking site LinkedIn appears to exist primarily to allow cultivation of these weak ties for job seeking. Malcolm Gladwell brought Granovetter's insight to a wide audience in his best-selling book *The Tipping Point*, where he observes, "Acquaintances, in short, represent a source of social power, and the more acquaintances you have the more powerful you are."[14] Gladwell uses this insight to identify "connectors," people with vast social networks, who he believes are a key to understanding how to successfully market and spread an idea. The success of Gladwell's popularization has made weak ties one of the best-known ideas from contemporary sociology.

Despite the apparent familiarity of the idea, it's worth returning to Granovetter's original paper to understand that not all weak ties are created equal. "The Strength of Weak Ties" begins with an analysis of sociograms, graphs of social networks. Granovetter is interested in bridge ties—"a line in a network which provides the only path between two points." These ties are important because they are the choke points in the flow of information and influence. Diffusion of ideas through a network depends on these bridge ties.

Granovetter's bridge ties have much in common with the bridge figures we're considering—they are part of two different social circles and can broker ideas between these networks—but bridge ties exist in social networks of people who share the same country and culture. The friend at a cocktail party who introduces you to a stranger who lives in the same building you do is a bridge tie.

It's difficult to ask an individual to identify the bridge ties in her social network. Answering the question requires knowledge you may not have—for example, that your friend Jane is well connected to a group of Latvian jugglers and could bridge between your social network and theirs. Because sociologists cannot easily study bridge ties through their usual survey methods, Granovetter proposes they study weak ties instead. His logic? Strong ties— ties between people who confide in each other, who see each other at least weekly—are never bridge ties. Here Granovetter relies on Georg Simmel's work on closure. If I'm close friends with Jim and with Jane, Simmel postulates, the two will feel intense social pressure to become friends. This is what explains why all the West African students at the college Wimmer and Lewis studied in their Facebook experiment are friends—it would be impolite not to be.

Closure is such a powerful effect, Granovetter believes, that he gives a special name to the situation in which I have strong ties to Jim and to Jane, and they have no ties to each other: "the forbidden triad." Because it's "forbidden" for two of my close friends to be disconnected from each other, strong ties don't serve as bridge ties; if you and I are closely tied, I'm likely to already know the people you are strongly linked to. Weak ties suffer no such restriction, though they are certainly not automatically bridges. What is important, rather, is that all bridges are weak ties. If we want to find the places in social networks where we could find connections to unexpected groups, we need to look beyond our closest friends and toward our weak social ties.

Granovetter's assumptions about strong ties may have been true in 1973 when he wrote the paper, but they are more questionable today. My wife, a congregational rabbi in our small town, is linked to hundreds of people in our geographic community and hundreds more through online discussions that let her interact with other congregational leaders around the world. Her strong ties in the geographic community may feel pressure to become friends with one another, but her online strong ties feel no pressure to know her local friends. This is not an uncommon pattern in today's age of social media: 50 percent of adult users of social media report a major reason for use is connecting with friends they've fallen out of touch with or are geographically distant from. Some 14 percent report using social media to connect with others who share an interest or hobby with them, maintaining ties that are geographically independent.[15] In an age of digitally mediated friendships, it's quite possible—and likely quite common—for strong ties to be bridge ties.[16]

Ultimately it's the bridge ties that matter even to Granovetter's analysis. He closes his paper with a look at two communities in Boston and their fights against urban renewal. The Italian community in the West End wasn't able to organize in opposition, while a similarly working-class community in Charlestown successfully opposed redevelopment. The difference, he concludes, is in the structure of friendships in those communities. West Enders belonged to tight cliques of friends, often people who'd grown up together. They worked outside the neighborhood and maintained close social ties to these friends in the community. Charlestown residents, by contrast, worked largely within the neighborhood, which gave them a chance to meet other Charlestown residents who weren't in their immediate circles of friends.

It's not that West Enders lacked weak ties. "It strains credulity to suppose that each person would not have known a great many others, so that there would have been some weak ties. The question is whether such ties were bridges." When it came time to organize, Charlestown residents had bridge ties—from work and voluntary organizations—within their neighborhood, while West Enders didn't. Granovetter speculates, "The more local bridges (per person?) in a community and the greater their degree, the more cohesive the community and the more capable of acting in consort."

It's not simply the number of acquaintances that represent power, as Gladwell posits. It's also their quality as bridges between different social networks. Lots of friends who have access to the same information and opportunities are less helpful than a few friends who can connect you to people and ideas outside your ordinary orbit.

Bridge Figures and Creativity

Weak ties may be able to help you find a job, especially if those ties are colleagues within an industry sector. Bridge ties provide a broader range of benefits. They're often the source of innovative and creative ideas.

Raytheon is the world's fifth-largest defense contractor, a multibillion-dollar company that builds everything from air traffic control systems to guided missiles. Their Patriot missiles featured prominently in the 1990–91 Persian Gulf War, and in response to sales requests from discerning governments around the world, Raytheon began expanding, acquiring four major defense-contracting businesses in the mid-1990s. Faced with the challenge of integrating these companies, Raytheon executives launched a close study of how ideas and best practices spread through organizations.

Ronald Burt, a sociologist and business school professor at the University of Chicago, was one of the thinkers they turned to. Burt worked with Raytheon from 2000 to 2003, serving as vice president of strategic learning, and testing his theories about social capital within the framework of a large and complicated enterprise. Burt believes that individuals who act as bridges between different social networks within a company "are at higher risk of having good ideas." These bridge figures often end up as "brokers" between disparate groups, sharing perspectives and different ways of thinking.

As a VP at Raytheon, Burt had freedom to design an unusual experiment, relying on extraordinary cooperation from corporate management. In 2001, he sent questionnaires to the 673 managers who ran the supply chain for the company. He asked each to document his or her connections to other people in the company with whom he or she discussed "supply-chain issues." A detailed sociogram of Raytheon's supply chain emerged. Burt calculated the "network constraint" of everyone in the organization: managers who spoke only to a densely connected network of co-workers, or who interacted primarily through hierarchies, were highly constrained, while those who connected with far-flung co-workers throughout the organization were unconstrained.

Raytheon, Burt discovered, did a pretty good job of rewarding managers who built bridges across "structural holes," gaps in the organization's structure that prevented people from talking to one another. The managers who were least constrained—the best bridges—were better paid than their peers, more likely to be promoted, and more likely to be evaluated as outstanding managers. They were also more likely to have good ideas.

Burt asked the study participants to share an idea about improving supply-chain processes at the company, then asked two senior executives in charge of supply chains at Raytheon to evaluate the ideas, stripped of all identifying information. Burt found small correlations between the best ideas and employee age (employees at the start and end of their careers had better ideas than those in the middle) and education (college-educated employees had better ideas than those with less education). But those effects were tiny in comparison with the correlations Burt found with social structure. "Even in the top ranks, people limited to a small circle of densely interconnected discussion partners were likely to have weak ideas for improving supply-chain operations,"[17] while those connected to a wide range of people were likelier to have better ideas, less likely to have their ideas dismissed, and more likely to discuss their ideas with others in the organization. The results were so pronounced that Burt titled his paper, simply, "Structural Holes and Good Ideas."

History celebrates the individual creativity of the solitary genius. We remember Edison, not the thousands of engineers who worked with him in Menlo Park. We can picture Einstein working alone at the patent office, but not "the Olympia Academy," a group of scholars he regularly met with when he lived in Bern.[18] Burt suggests that it might be time to let go of the idea that creativity is a function solely of personal genius. Good ideas, he argues, are a function of social structure as well: "People connected to groups beyond their own can expect to find themselves delivering valuable ideas, seeming to be gifted with creativity. This is not creativity born of genius. It is creativity as an import-export business."[19]

This import-export business works in multiple ways. At its simplest, brokers make their colleagues aware of the interests and challenges another group faces. Sometimes they're able to transfer best practices from one group to another. Higher levels of brokerage, Burt suggests, involve drawing analogies between groups, escaping the tendency to emphasize the differences between groups and instead recognize similarities. At the highest level, brokers offer synthesis of ideas between groups, novel solutions that combine thinking from different groups. In other words, they make curry.

It's worth remembering that the managers within Raytheon are bridging between divisions in the same American company. When Burt discusses cultural differences between groups, he's talking about differences between managers who purchase from outside contractors and those who purchase from other departments within Raytheon. And yet, despite the apparently low cultural barriers, Raytheon has had a very hard time implementing the innovations suggested by high-level bridge figures. Burt visited Raytheon a year after his survey and asked another senior executive to look at the top hundred ideas generated in his research—on eighty-four of the ideas, no steps had been taken toward implementation. "There was a brokerage advantage in producing ideas, and company systems were working correctly to reward brokers . . . but the potential value for integrating operations across the company was dissipated in the distribution of ideas." Bridge figures could identify opportunities for Raytheon, but the corporation often wasn't able to spread and adopt those new ideas.[20]

The Foreign Correspondent as Bridge

If the "import-export" of ideas within a single company is complicated, it's vastly more complex when we consider the challenges involved in encountering ideas from around the world. The bridge figure seeking fusion between the best features of two worlds may be brokering connections between people who speak different languages or practice different faiths. While people who work for the same company have (at least in theory) the common mission of that corporation's success, people in different parts of the world may be working toward different, and sometimes competitive, goals. The local biases of media attention and our personal homophily mean we're likely to encounter far more ideas from our near neighbors than from other parts of the world.

For years, foreign correspondents have worked as bridge figures, telling their audiences about the events, challenges, and attitudes of the lands they're living in. The first foreign correspondents were literally letter writers, corresponding with friends back home. They shared news from local newspapers and from what they saw in far-flung cities, and their letters were printed as dispatches in early newspapers. The hunger for international news in the eighteenth and nineteenth centuries meant that American newspaper editors sent copyboys to meet ships as they pulled into harbor, hoping to scoop their competitors in providing the latest news from Paris, London, and Amsterdam.

The arrival of the telegraph transformed the established model of foreign correspondence. In 1850, Paul Julius Reuter left a job with Agence France-Presse in Paris and moved to Aachen in the kingdom of Prussia, close to the borders with the Netherlands and Belgium. Using the newly completed Aachen–Berlin telegraph, and a set of homing pigeons, he began reporting business news from Brussels to readers in Berlin and transformed expectations about the speed at which we encounter international news. The completion of the telegraph line connecting the United States and Britain in 1857 began an age in which news could travel around the globe far faster than human beings could.

Reuter made his fortune brokering international information that required little context or interpretation: share prices from European stock markets, dramatic news headlines like the assassination of President Abraham Lincoln. More complex stories required context and interpretation for audiences at home. William Howard Russell's lengthy dispatches from the Crimean War (1853–56) for the *Times* of London established a model that informed foreign correspondence for subsequent centuries. His missives weren't intended to offer breaking news. They arrived weeks later than shorter reports of troop movements and battles fought.

Instead, Russell's dispatches painted a vivid picture of conditions on the ground for combatants and civilians, connecting British readers to a conflict fought far from their shores. Historians credit the power of Russell's dispatches with inspiring Samuel Morton Peto and other British railroad contractors to build a railway line to supply soldiers in the siege of Sevastopol, now viewed as a turning point in the war. And Florence Nightingale described Russell's writing as her inspiration to bring a group of nurses to tend to the Crimean War wounded, greatly reducing the death rate in field hospitals and reshaping models for contemporary nursing.

The need for connection and context remains in the age of "parachute reporting," where journalists armed with satellite uplinks and video cameras were able to report from earthquake-ravaged Port au Prince within hours of the devastating 2010 Haitian earthquake. But most of those reporters spoke no Kreyòl and knew little about Haiti before the quake. Much of the best reporting came from journalists who'd lived and worked in Haiti before the quake, writing for US and European newspapers, like Jacqueline Charles of the *Miami Herald*. Charles, a Haitian and Turks Islander, began working for the *Herald* as a high school intern in 1986. By the time the 2010 quake struck, Charles had covered earlier Haitian disasters that went largely unnoticed in the global media, like a set of tropical storms that destroyed the town of Cabaret in 2008.

A native of the Caribbean who studied journalism in North Carolina, Charles is exactly the sort of bridge figure we would expect to be effective as a foreign correspondent. Her knowledge of US audiences and Haitian realities allows her to explain events in terms her audience will understand. Many news organizations have not yet made the shift from the William Russell model—an Irishman reporting on Crimea for British audiences—to a Charles model—a Haitian educated in the United States, reporting on Haiti to an audience in Miami. Traditionally, foreign correspondents have been travelers from abroad, reporting news to audiences at home, not locals writing for an international audience.

Solana Larsen, the Danish Puerto Rican managing editor of Global Voices, picked a fight at a US journalism conference by announcing her hope that foreign correspondents would become a thing of the past and that media outlets would become more reliant on local reporters, helping them contextualize their stories for global audiences. Several journalists called her suggestion naïve and irresponsible. But Richard Sambrook, then the head of global news for the BBC, spoke in her defense. The BBC, he explained, was moving away from parachute journalism and toward a future where hundreds of local stringers reported for British and international audiences.

The key is context. Without a clear understanding of what audiences know and don't know, stories from different parts of the world can be completely incomprehensible.

We featured a cartoon on Global Voices in August 2010 that depicted the Russian leader Vladimir Putin talking on a mobile phone. We translated the Russian dialogue: "Abramovich? Hello! Listen, do you have a rynda on your yacht? See, the thing is . . . You have to return it." Unless you follow Russian news very closely, you probably need a translation of the translation.

In the summer of 2010, western Russia suffered a major heat wave and a series of hundreds of wildfires. The fires destroyed the homes and property of thousands of rural residents, and the smog from the smoke, combined with the intense heat, led to the death of many elderly and infirm city dwellers. Roughly 700 Muscovites died per day in early August 2010, twice the normal death rate. The insurance firm Munich Re estimates that 56,000 Russians died of direct and indirect effects of the fires.

The Russian government, particularly Prime Minister Vladimir Putin, came under intense criticism for its perceived inaction in the face of the fires. A Russian blogger, "top-lap," complained in a post that his village was better prepared to fight these fires before the fall of communism:

> Do you know why we're burning? Because it is a fuck up. In my village under communists—who are being criticized by everyone now—there were three fire ponds, and a rynda that people would ring in case of fire and—oh, miracle—fire truck, one for three villages but at least there was a fire truck. And then democrats came and that is when a fuckup started. They leveled the fire ponds with the ground and sold that ground for construction projects. They did something to the fire truck, maybe aliens stole it, and the rynda was replaced (fucking modernization) with a telephone that doesn't work because they forgot to connect it to the line.

Top-lap ended his diatribe with this demand: "Give me my fucking rynda back and take your fucking telephone." His post was spread throughout the Russian Internet by Aleksey Venediktov, the editor in chief of Russia's most influential opposition radio station, Echo Moskvy. Remarkably, Putin responded to the post, explaining that unprecedented high temperatures had caused the fires and that the government was working hard to respond to them. His message closed with this assurance: "If you provide your address, you can receive your rynda from your governor immediately."[21]

Unsurprisingly, Russian bloggers had a field day with Putin's comments, and "rynda"— previously an archaic and rarely used term for a small bell—emerged as a symbol of Russian dysfunction in an age of crony capitalism. In the cartoon, Putin is calling

Roman Abramovich, the billionaire owner of the Chelsea Football Club. Abramovich, a well-known and powerful oligarch, profited from the collapse of the Soviet Union by purchasing valuable state-owned assets, like the oil company Sibneft, at fire-sale prices. Asked to bring top-lap's rynda back, Putin is forced to call the oligarchs who benefited from the end of the communist era.

The author who took on the challenge of explaining the cartoon, bridging between Russian media and the global Internet, was Vadim Isakov, an Uzbek blogger and journalist who, coincidentally, trained at the same US journalism school as Jacqueline Charles. He has worked as a Central Asian correspondent for Agence France-Presse and as a media trainer in Uzbekistan and now teaches communications at Ithaca College in New York. In other words, he's a bridge figure able to identify the features of the story that would appeal to a Global Voices audience—the use of new media to confront authority, humor, the spread of a meme online—and the background necessary to understand and appreciate the cartoon.

Bridging in a Digital Age

For people like Vadim who are able to bridge Russian Internet humor for international audiences, the Internet provides a rich set of tools. Used well, they can give bridge figures superpowers.

In my explanation of the Putin/Rynda cartoon (stolen largely from Vadim), I've opened by using a standard journalistic technique, the "nut graf." The nut graf is a quick summary of events that provide context for a story. In a feature story, where an anecdote is used to illustrate a larger event, the nut graf supplies the context; in my example above, a paragraph about the Russian fires offers background for the reader to understand the importance of the top-lap anecdote. In news stories, the nut graf provides context to a recent development: in a story about a vote on an immigration bill in Congress, the nut graf might summarize debates over immigration during the past few years.

Like many journalistic inventions, the nut graf is an elegant adaptation to the limitations of the form. Space is scarce within the pages of a newspaper, and the same story needs to inform both someone who's following a story obsessively and someone watching it casually. Those limitations aren't true for online media; a nut graf can expand, accordionlike, into an "explainer." The journalism professor Jay Rosen unpacks the term: "An explainer is a special feature that does not provide the latest news or update you on a story. Rather, it addresses a gap in your understanding: the lack of essential background knowledge, such that items in the news don't make sense, fail to register as important or add to the feeling of being overwhelmed."

To illustrate the explainer, Rosen points to "The Giant Pool of Money," an hourlong documentary produced by the radio program *This American Life*. The documentary dives deep into the mortgage crisis that rocked global financial markets in 2008, and it was the most popular ever produced by *This American Life*. For Rosen, it had another effect: "I became a customer for ongoing news about the mortgage mess and the credit crisis that developed from it. (How one caused the other was explained in the program's conclusion.) 'Twas a successful act of explanation that put me in the market for information."[22]

Without context, a news story can be overwhelming and incomprehensible. It implicitly sends a message that we don't know enough about an issue to understand the story's importance. Participatory media—blog posts from unfamiliar countries, tweets from protests or war zones—are even harder to understand. If we can make rich, compelling explainers, timelines, and backgrounders, Rosen argues, we can expand the audience for news stories that often get ignored.

Sometimes the context for a story isn't enough. Even in translation, stories from other parts of the world can require glossaries to make them understandable. This is particularly the case in stories about the Internet in China, where fears of government censorship encourage online authors to use sarcasm and humor to get their points across.

The blog chinaSMACK offers English speakers an irreverent look at the topics Chinese people are talking about in online forums, in dorm rooms, and around dinner tables, with an emphasis on the shocking, controversial, and weird. Few details are known about the site's editor, who goes by the pseudonym "Fauna." She has told reporters in email interviews that she's female, from Shanghai, and started translating posts from online forums in 2008 as a way to refine her English skills.[23] The site she now manages is viewed by roughly a quarter million people a month, mostly in North America and Europe. They visit chinaSMACK in part because the site does such an effective job of contextualizing the strange videos and stories posted on the site.

A photo essay about migrant laborers in Guiyang who make their living from picking through garbage, posted on the Netease web portal, is translated on chinaSMACK, along with a sampling of the comments posted on Netease.[24] Most comments express sympathy for the poor, suggesting campaigns to raise funds to pay for the education of their children. Others have more complicated and nuanced meanings:

These children may spend their entire lives without being able to ride the high-speed trains, drink Maotai, nor will they be able to take out money to donate to the Red Cross Society. Their hearts are indeed pure, not feeling that what happens to them is unfair, quietly accepting their reality, while us bystanders can only express indignation towards the unfairness in this society. . . . Those in support please ding this more.

ChinaSMACK has helpfully hyperlinked the terms "high-speed trains," "Maotai," and "Red Cross Society," as well as the term "ding." The first three lead to collections of articles about recent scandals in Chinese society: the crash of high-speed trains on the Ningbo–Wenzhou line; a scandal regarding Sinopec's corporate spending on expensive rice wine; and stories about a wealthy twenty-year-old woman, Guo Meimei, rumored to be central to a corruption case in the Chinese Red Cross. "Ding" links to a glossary entry, explaining the term's similarity to clicking "like" on Facebook to promote a story to a wider audience.

Decoded and contextualized, the pseudonymous comment is a masterpiece of snark, inviting Chinese netizens to consider which issues are worth getting outraged over and which should be dismissed as trivial. With the explainer embedded into the article, a story about scavenging in a garbage dump becomes an invitation to follow links to other stories where terms have emerged, from articles about Chinese support of charities in Africa to government spending on the Shenzhen Universiade international sports competition.

The ability to traverse a set of hyperlinks is only one way in which the Internet is especially fertile ground for contextual bridging. The rise of photo- and video-sharing services allow bridge figures not just to write about the places, people, and events essential to understanding local context but also to show them. When I want to explain the entrepreneurial culture of West Africa, I no longer have to paint scenes of people carrying trade goods on their heads and hawking them to passersby. I can search for "Ghana OR Nigeria AND market" on Flickr and lead a reader through an impromptu slide show of market cultures. None of the images in the streams are mine; they're posted by Ghanaians, Nigerians, and travelers to those countries and shared under Creative Commons licenses, allowing me to republish the images and credit the photographers without paying licensing fees.

Participatory media's power for bridging extends beyond A/V

show-and-tell. The people who write for Global Voices, bridging between the communities they report on and audiences around the world, tend to be fanatical users of social-networking tools like Facebook and Twitter. If I find myself intrigued by Vadim Isakov's take on Russian social media, I can "friend" him on Twitter and see what other stories he's onto. Getting information about his likes and interests aside from the stories he writes on Global Voices helps him become more three-dimensional to me. When I hear stories about Uzbekistan, I can connect them to a person I "know" through social media, though I've never met him in person. I move a step closer to solving "the caring problem," and Vadim becomes a more effective bridge for me because I understand his interests and foci and can see what he might be emphasizing or missing in his accounts. I also get the side benefit of catching glimpses of global media through the eyes of an Uzbek professor, who's likely to pay attention to some stories I might have missed.

Being able to relate personally to a bridge figure can be a complicated experience, especially when you're building bridges between people who've been at war.

Like many Armenians, the activist Onnik Krikorian didn't know any Azeris growing up. Although tensions between Armenians and Azeris date back to pogroms at the turn of the twentieth century, Krikorian's generation remembers the Nagorno-Karabakh War, where from 1988 to 1994 the countries fought over a disputed province. The tensions around this stalled conflict create mistrust between Azeris and Armenians; in a recent survey, members of both groups overwhelmingly condemned the idea of making friends with people from "the other side."

The problem, Krikorian believes, is that there are so few spaces in the physical world where Armenians and Azeris actually interact with one another. On his blog, he writes movingly about teahouses in Tbilisi, the Georgian capital, where Azeri singers perform Armenian songs for a clientele that spans the Caucasus.

Since most youth in Armenia and Azerbaijan aren't able to travel or study abroad, Krikorian is encouraging connection in the one space they have in common: Facebook. "Two years ago, an Armenian befriending an Azeri on Facebook would have been unthinkable," Krikorian says. Through a set of online workshops, video chats between members of the two communities, and endless online and off-line diplomacy, Krikorian has helped encourage contact between youth in the two countries. "We're seeing simple things—an Azeri wishing an Armenian friend happy birthday. But this would have been impossible until very recently."

Zamira Abbasova, an ethnic Azeri whose family fled Armenia when she was four years old, encountered dozens of Armenians as a student in the United States, but it took her years to "check the expiration date on her hatred" and begin making friends.

"The difficult step is often to get people to acknowledge and display the contacts that are taking place. When Zamira posted [on Facebook] about losing her hatred for Armenians, she was flooded with abusive email. Another participant received a death threat, complete with a picture of a bloody corpse," Krikorian explains. While the Internet enables contact between people who've historically been in conflict, it doesn't guarantee that they'll interact or that, when they do, it will be positive. What Krikorian and friends are doing, attempting to build bridges online, instead of in Georgian coffee shops, is challenging and sometimes dangerous. Their form of bridging requires a willingness to persist in the face of criticism, resistance, and threats.

Human Libraries

I was recently in Nairobi, Kenya, researching the use of electric power in poor neighborhoods. There are many good guidebooks to Kenya, including ones that give overviews of "slum tours" to neighborhoods like Kibera. But I've yet to find a guidebook or website that could tell me how to visit dozens of shops in poor neighborhoods and ask their owners whether they used grid or generator power. For some questions, you don't need an answer, you need a guide.

If bridge figures are key to crossing contexts in a connected world, the problem of finding an appropriate guide remains a tricky task. I got lucky: one of my students was staying with a friend who manages an arts center that works with youth in Nairobi's poorest neighborhoods. She found a musician from the Baba Dogo neighborhood who led my students and me around, and translated our nosy questions into ones shopkeepers were willing to answer.

To find a guide, I needed someone who understood my research questions and who knew experts on neighborhoods in Nairobi. A new wave of Internet services is trying to answer questions in a similar way, posting your questions to a set of people, and trying to find expert guides.

Prior to a recent trip to Adelaide, Australia, I posted a question to Härnu, a new service named for the Swedish words for "here" and "now." I asked what websites I could read to understand local politics in the city before talking to government officials, and within a few hours, I had half a dozen suggestions. My question had been posted to a virtual map, pinned to the city of Adelaide. Anyone was welcome to field my question, but the answers came from Härnu users who followed questions posted to South Australia. I've begun following the site, trying to field inquiries posted to western Massachusetts and West Africa, two areas I'm knowledgeable about.

Had I wanted an "expert" answer to my question, I could have turned to Quora, where tech entrepreneurs like Steve Case and Marc Andriessen have answered questions and where Marc Zuckerberg, the Facebook founder, has posted to ask what companies Facebook should acquire. Whether they're staffed by Silicon Valley royalty or helpful South Australians, the services run on the same basic principle: they match people with questions to individuals with answers, and rank individual expertise on different topics on the basis of how satisfying those answers were to users.

Other sites attempt to identify experts in terms of their social media influence on particular topics. Klout tracks the posts individuals make to Twitter and Facebook and looks to see how widely that information spreads, giving users a measure of their influence, or "Klout." For users Klout knows a lot about, it suggests topics they're influential (and presumably knowledgeable) about. Klout thinks I'm influential about entrepreneurship and Africa (perhaps) as well as academics and prison (less likely). It's easy to see how this service could turn into a search engine to identify experts on key topics (or people whom PR people should flood with press releases, hoping to influence technology "influentials").

The experts who make these services possible are a type of bridge. They connect the general public with specialized knowledge—the sites of Hong Kong, the politics of Ghana—tackling many of the same challenges of context. The sites suggest a future where the Internet connects not just people to information but people to knowledgeable people, a reality in which bridging, contextualizing, and explaining will need to move to the center of online interactions. This is a future in which those best able to bridge will be some of the most powerful in creating and sharing knowledge.

This idea of connecting people to people isn't a new one. Socrates taught through dialogue, not through written texts, and Plato famously observed that books, unlike people, always offer the same answers. In response to issues of urban violence in Copenhagen in the early 1990s, a group of activists set up a "human library" of "living books": people who could be "checked out" for a brief conversation by others who wanted to speak with a person from a different background, to confront and overcome prejudices. The idea has spread to communities in Australia and Canada, where human libraries have expanded to include experts on community history, as well as representatives of different ethnic and religious communities.[25]

The library in my hometown, a college town in rural Massachusetts, recently held a human library day, and I attended, planning to "check out" a young undergraduate from Ghana, a country I've regularly visited since the early 1990s. I hoped to introduce myself and offer my services as a bridge to our local community for him, but I didn't get the chance. The program was so popular, the Ghanaian student was booked solid, explaining West Africa to New Englanders, and all other "living books" were checked out for the day within the first hour of the event.

Not all projects take the notion of human libraries quite so literally. They look for people who can act as guides to realms of knowledge outside of established educational institutions. Achal Prabhala, an Indian intellectual-property activist and adviser to Wikimedia, is trying to get the vast online encyclopedia to acknowledge the complex truth that "people are knowledge." In a documentary film funded by the Wikimedia Foundation, Prabhala and his collaborators explore the challenges Wikipedia has had in incorporating knowledge from communities in India and Africa. Much of the important local knowledge isn't in print; it's in recipes known by women in villages, in stories told by community elders, or in games played by generations of schoolchildren. Wikipedia's rules on sourcing—banning original research as a citation for an article and demanding existing citations in print or online—don't work in these cases. Prabhala proposes that Wikipedians start documenting knowledge from these communities through video and audio interviews, both creating a body of indexed knowledge that didn't exist previously and bridging the individuals who have this knowledge and the rest of the world.[26]

In this case, both Prabhala and the elders he's working with are acting as guides. The elders are able to connect him with experts on undocumented cultures, and Prabhala is able to decipher the complexities of Wikipedia, helping unlock their knowledge for a global audience.

Human libraries and Prabhala's expansion of Wikipedia remind us that the Internet is far from the only space where we might encounter unexpected knowledge. The Internet is special in that it makes it trivially easy to encounter people and information from other parts of the world, if we choose to. But we don't always encounter guides as skilled as Prabhala, or situations as carefully configured as a human library. As we think about rewiring the Internet to encourage connection, we need to think about how to build spaces and institutions that help bridge figures and xenophiles.

Xenophiles

For bridge figures to be effective, someone has to cross the bridges they build. If bridge builders invite people to explore and understand different cultures, xenophiles enthusiastically accept that invitation.

Dhani Jones, recently retired as a middle linebacker for the Cincinnati Bengals, is a xenophile. The spring of 2010 found him playing water polo in Croatia, tossing the caber in Scotland, and learning Laamb—traditional wrestling—from a massive wrestler nicknamed "Bombardier" on the beaches of Dakar, Senegal. The premise of his television show, *Dhani Tackles the Globe*, was simple: he would visit a country for a week, train with top local athletes, and compete in a sport he'd never played before.

It requires extreme physical talent to step into the ring with a professional Thai kickboxer after a week of training and survive the experience, but Jones's most impressive traits allow him to build connections with fellow athletes and nearly everyone he meets on the show. He projects a sense of openness, good humor, and approachability that lets people reach out to him and celebrate the best features of their cultures.

A trip to Russia to study the martial art Sambo finds Dhani standing on a bridge in St. Petersburg, attempting to drop coins onto a narrow ledge above the water—a practice said to give good luck to those dexterous enough to land a coin. An elderly man passes by, and in a mix of smiles and hand gestures, Dhani enlists his help. The two, the old man's shaky hand guided by Dhani's massive one, drop coin after coin onto the ledge, high five, and hug. It may not be as marketable a set of skills as tackling a running back in the open field, but it's damned impressive, as anyone who has tried to build international friendships as a tourist knows.

Dhani's not a bridge figure; he's just well traveled. In his biography, he traces his love of travel to a trip to Paris and East Africa with his parents when he was four. But he was born and raised in the United States, and while he's passionate about his time in Senegal and Singapore, he's not the right person to explain the intricacies of those cultures to the wider world. He's in love with the diversity and breadth of human experience, and he's willing to cross bridges to get to that wider world.

It's no accident that passions make people encounter other cultures: Paul Simon's fascination with mbaqanga music, Dhani Jones's excitement about British rugby and Jamaican cricket. The television network that bought twenty episodes of Dhani's show, the Travel Channel, is best known for airing No Reservations, a show in which the chef Anthony Bourdain eats his way around the world, usually accompanied by local chefs he admires and befriends. (Bourdain, in turn, acts as a bridge, rather than as a xenophile, when he invites us into the secret society of restaurant chefs in books like Kitchen Confidential, his memoir about his life in professional kitchens.)

Passion translates well across borders, and a shared passion— particularly a passion put into practice—leads to interaction.

Or as Dhani notes, reflecting on his motivations as an athlete, "I think I play sports because it allows me to connect. . . . I've built deep relationships, through competitive sport, with people who have never talked to a black man, let alone an NFL player."[27]

If digital media make it easier for bridge figures to put their cultures in context, they completely transform the life of a xenophile. In the late 1980s and early 1990s, after Paul Simon had helped bring a few African records into American record stores, I was obsessed with Afrobeat and Afrojuju, trying to learn all I could about Nigerian musicians like Shina Peters. This involved taking the train to New York City, interrogating baffled record store clerks, and eventually discovering that Afro-Caribbean grocery stores in the South Bronx were the best hunting grounds. A quick Google search for Shina Peters today turns up a well-referenced biography and discography, hard-to-find albums on sale on eBay, and dozens of concert videos, footage I would have killed for twenty years back.

Musical experimentation across cultures is no longer limited to artists whose record companies can broker introductions to the best musicians in another nation. The "techno-musicologist" Wayne Marshall studies "nu-whirled music," the strange cultural hybrids that are possible in an age where cultural influences are a YouTube search away. In a lecture at Harvard, he traced an LA street dance style—jerkin'—where young dancers dressed in neon shirts, tight jeans, and colorful Chuck Taylors strike angular poses to a synthesized beat. The next video shows jerkin' in Panama, where a seminal jerkin' track—New Boyz, "You're a Jerk"—has been remixed with a Spanish-language rap on top. The Panamanian kids have cut a video as well, using footage from the New Boyz video and scenes of Panamanian kids in their best jerkin' clothes. Two videos later, and jerkin' has moved through the Dominican Republic, getting remixed with "dem bow," a Dominican variant of Jamaican reggaeton and emerging as "jerkbow." And now Dominiyorcan kids in New York are showing off their jerkin' moves, dressed in LA neon, in playgrounds with snow on the ground.[28]

The next generation of musical xenophiles are making art from this fluid world of "global bass music" or in a tongue-in-cheek term Marshall coined, "global ghettotech." One of the stars of the space is Diplo, known to his parents as Wesley Pentz. Raised in Mississippi, he developed an infatuation with the dance music style called "Miami Bass" that led him to become a DJ and a musical explorer, mixing global dance music in DJ parties in Philadelphia. Diplo quickly became known for his love and knowledge of "baile funk," a remix of Miami bass for the favelas of Rio.[29]

Baile funk[30] was largely unknown outside Brazil until Diplo produced "Bucky Done Gun" for the Sri-Lankan British singer Maya Arulpragasam, better known as M.I.A. Released on an influential mix tape, "Piracy Funds Terrorism," the song samples heavily from Deize Tigrona's baile funk song "Injeção," which, in turn, samples the horn lick from the *Rocky* movie theme song. The influences in this popular track are fascinating to trace: a three-decade journey back from Diplo to baile funk, to Miami bass, to Detroit techno, to the early American electro hip-hop of Afrika Bambaataa to the German synthesizer pioneers Kraftwerk.[31]

Both Diplo and M.I.A. have taken flak from critics and fellow musicians for their working method: taking elements of different musical cultures and mixing them into new, hybrid forms. Is this appreciation or appropriation?

Asked by a Brazilian reporter whether his work runs the risk of trivializing the music he's celebrating, reducing it to "the flavor of the month," Diplo responded,

My job is just a DJ/performer, not a sociologist—and I do a good job at it, collecting and introducing fresh sounds to my audience (it goes back to the old days when hip hop DJs championed new music and kept it secret from each other by covering the labels on records—to have an upper hand on other DJs). But since my livelihood depends in some way to these subcultures existing, I have set up some things to help them develop.[32]

The "things" Diplo refers to include a documentary and a program that works with indigenous Australian youth to produce dance music.

Diplo clearly sees himself as an ambassador for global bass music. Whether or not he always gets it right, it seems that Diplo is taking seriously the cosmopolitan notion of responsibility to others, not just celebrating the musical artifacts he finds in his journeys online and around the world.

This sense of responsibility separates the xenophile from the appropriator, the collaborator from the musical tourist. Discovering that sense of obligation to others can take time.

In 2003, the American videogame designer Matt Harding left his job in Australia and began traveling, making short videos of himself doing the same goofy dance in different places around the world. He edited dozens of clips into a strange travelogue—he's in the center of every frame, dancing badly, a constant against a changing backdrop of remarkable sights. Matt put the video, titled *Where the Hell Is Matt?*, on his website and sent a link to a few friends. In a few months after the release of his first video, which became a viral hit, Matt had one of the world's strangest jobs; he was paid to wander around the world making funny videos, which he subsequently released to the web in 2006 and again in 2008.

There's a noticeable shift between Matt's second and third videos. In the first two videos, he's a lone figure dancing in front of remarkable backgrounds. A minute into the third video, dancers rush into the frame, crowding Matt out and turning the video into a series of joyful mobs dancing in public in places ranging from Madrid to Madagascar. In thinking about what to do for his

third video, Matt recalled, "I was beginning to realize that dancing in front of exotic backgrounds was a thin gimmick. I'd found
what I should've been doing all along. I should have been dancing
with other people."[33]

If the first two videos are stories about one man's remarkable
adventure around the globe, the third is a story about humanity
and the ways in which people connect with one another. Matt
and his girlfriend organized these shoots via email, inviting fans
to come and dance with Matt as they came through a city. In
places where Matt's online celebrity hadn't reached the general
public, like Sana'a, Yemen, Matt's dancers are neighborhood
children. One of the most touching moments in the 2008 video
is a cut from Matt dancing with a massive crowd in Tel Aviv to
him dancing with a small group of Palestinian children in an alley
in East Jerusalem, a transition included at the insistence of an
Israeli participant who told Matt, "Put them together. They must
be side by side, one right after the other."[34]

There's another difference between Matt's second and third
videos: the music. The soundtrack for Matt's original video was
Deep Forest's "Sweet Lullaby," a song by French electronic musicians with a long and controversial history. Describing themselves
as "sound reporters," the members of Deep Forest claim that
their albums represent the voices of African pygmies: the lyrics
to their debut album's first song describe men and women living
in the jungle and suggest they are both our past and possibly our
future.

Perhaps little men in the jungle are our future, but they're not
the ones singing on "Sweet Lullaby." The track is built around a
lullaby called "Rorogwela," sung by a woman named Afunakwa,
recorded in the Solomon Islands, roughly half a planet away from
Central Africa. The melody was recorded by the legendary ethnomusicologist Dr. Hugo Zemp, and when Deep Forest asked his
permission to use the recording, he refused. Deep Forest used the

sample anyway, Zemp wrote angry academic articles about cultural appropriation, and the scandal over the vocal didn't stop the album from selling millions of copies.[35] Near as I can tell, no one attempted to contact Afunakwa and share royalties with her.

Matt Harding heard the story about Afunakwa, and he decided to handle the music differently for his third video. He commissioned an original orchestral piece by the composer Garry Schyman, with a Begali vocal sung by Palbasha Siddique, adapted from a poem by Rabindranath Tagore. Harding's motives here weren't entirely idealistic—his videos were so popular that he now needed to weigh the risk of a copyright suit from Deep Forest. But his next steps only make sense in the context of responsible xenophilia.

One of the most energetic shots in the 2008 video shows Harding dancing with a room full of ecstatic children in Auki, capital of the Solomon Islands. Matt paused his around-the-world trip for the third video to shoot a short documentary, titled *Where the Hell Is Afunakwa?* It has been seen orders of magnitude less often than his dancing videos, but it marks an attempt to close the loop of cultural borrowing, explaining the story behind "Sweet Lullaby," interviewing Afunakwa's descendants and hearing about her death in 1998.

In 2011, Harding returned to the Solomon Islands to find Afunakwa's son, Jack. After an epic trip involving a flatbed truck and a motorboat, Harding met Jack and worked with his sons to set up an ad hoc committee that would accept a share of profits from the videos and use the proceeds to pay medical and school fees for Afunakwa's descendants. Before he left, Harding wrote on his blog, "I stopped by at the mission, found the headmaster, and paid the annual school fees for all of Afunakwa's descendants who are of age. It cost slightly more than my monthly cable bill."

Harding now regularly emails with Godfrey, Afunakwa's grandson, to inquire about the community's needs and to coordinate wire transfers of money from the United States to Afunakwa's family. A journey that began as a tour of beautiful places around the world has turned into a righting of a wrong, and a bond of responsibility between a xenophile Internet celebrity and a village in the South Pacific.[36]

In 2012, Harding released a fourth video, which suggests his journey from goofball to xenophile continues. The video opens with Matt taking dancing lessons in the streets of Kigali and Seville. As the music swells, Matt's no longer performing his silly dance, but following the gestures of men in dishdasha in the Saudi desert and shaking his hips with a crowd in Port-au-Prince. As he attempts to waltz with an elegantly gowned woman in Pyongyang, it becomes clear that Matt still dances badly, but now he has the world as a dance teacher. As crowds in Cairo, Tallinn, Helsinki, and Hong Kong thrust their arms into the air, echoing one another's gestures, the logic of the video becomes clear: Harding is trying to get his friends all around the world to dance together.

The hard challenges we face as a species can't be solved by even the most talented dance instructor. But Harding's experiment is a first step in solving some of the thorniest problems in global communication. If we accept Ronald Burt's invitation to look for creative solutions at structural holes, gaps in our knowledge about the world, we're led to look for ideas and inspirations from other cultures. But we quickly bump into Joi Ito's "caring problem"— without a way to build personal connections to people from other parts of the world, it's hard for us to take their perspectives and insights seriously.

Encountering the world through shared interests is a shortcut to encounters that connect us with other human beings. It's difficult for us to rise to the challenge of paying attention to news that affects people we don't know in places we've never been— finding a shared interest is a first step toward finding inspiration and insight in the unfamiliar. That step, even as a sloppy dance step, is a move toward problem solving that incorporates diverse and complementary ways of thinking.

Important as xenophiles and bridge figures are in exchanging ideas, they can't correct the shortcoming of our media—and our views of the world—by themselves. To follow the trails they are blazing for us, we can't rely purely on our desire to encounter a broader world. We need to harness a powerful, and poorly understood, path to discovery: serendipity.

Chapter 7

SERENDIPITY AND
THE CITY

WILLIAM GIBSON'S 1984 BOOK *NEUROMANCER* OFFERED A vision of the Internet as a physical space, a vast, colorful city of buildings representing the computer servers owned by global corporations. Only "console cowboys," Gibson's hacker heroes, could access this imaginary space, "jacking in" to the Internet through custom hardware, but to them it was as vast, intricate, and real as any city in the world.

Neal Stephenson's *Snow Crash* followed eight years later, and invited us to think of the Internet as "the Metaverse," an immersive 3-D world populated by digital avatars, controlled by their goggle- and glove-wearing users. Stephenson's Metaverse is a vast, black, mostly empty planet. The main space that's inhabited is "the Street," a linear city that circles the globe, where users come to see and be seen, to rub elbows and bump shoulders in a virtual environment.

Why is it so tempting to think of the Internet as a city? There's no reason data can't be visualized as a forest or a sea of bits, an endless desktop stacked with documents, Borges's infinite library. Cities are an insane way to visualize data—why would we force people into close contact when we're building "spaces" that can be infinite in scale?

To understanding the appeal of the digital city, we need to consider the charms of the physical city.

Cities and Choice

Makoko has been referred to as "Nigeria's Venice," with houses, stores, and churches linked by plank bridges that cross above the waters of Lagos Lagoon. The wooden pirogues that ply the waters between buildings in this dense slum aren't carrying tourists and singing gondoliers. They carry fish from the lagoon and boards from the sawmills that line the lagoon. They bring goods to market and children to schools.

What began as a small fishing village on the outskirts of Nigeria's commercial capital, Lagos, has been transformed into one of the densest neighborhoods of a notoriously crowded city. Estimates vary, but most observers believe that at least 100,000 Lagosians live in a neighborhood that now extends half a mile into Lagos Lagoon. What brings people to this neighborhood is not prestige—it's considered one of Lagos's most dangerous neighborhoods—or the waterfront views. It's not the amenities— there's no running water, and outhouses dump directly into the lagoon. Electricity, pirated from lines on shore, is in short supply and dangerous when it's on. Cholera and other diseases are common. In July 2012, the governor of Lagos state ordered the destruction of thousands of illegally built dwellings, and knocked down dozens, leaving their owners homeless.[1]

People come to Makoko because Lagos is growing, and there's nowhere else to go. An estimated 275,000 people move to Lagos each year, roughly the same number who lived in the city in 1950, and by some estimates, the city of 7.9 million is now more populous than London. The small islands that serve as the commercial and state capital are thoroughly developed, and nightmarish traffic jams make living in the outskirts less appealing. Some newcomers to Makoko are literally making their own piece of Lagos, alternating layers of garbage diverted from landfills (going price is about fifty cents a truckload) and sawdust from nearby saw-mills (which helps mask the stench) and topping the reclaimed land with sand, and then structures made from wood and corru-gated roofing.

The residents of Makoko are part of a global trend toward urbanization. As of 2008, the majority of the world's population lives in cities. In highly developed countries (the membership of the OECD), the figure is 77 percent, while in the least developed countries (as classified by the UN), 29 percent of people live in cities. It's an oversimplification, but one way to think about eco-nomic development in the nineteenth and twentieth centuries is to emphasize the shift from a rural population, supported by subsistence agriculture, to an urban population engaged in man-ufacturing and service industries, fed by a small percentage of the population that remains focused on farming. As developing nations industrialize, the shift continues and there's a steady rural-to-urban migration.[2]

In 1800, only 3 percent of the world's population lived in cities, many in European capitals like London and Amsterdam. Even so, those societies had rural majorities: roughly 80 percent in England, 75 percent in the Netherlands. A century later, 14 per-cent of the world's population resided in cities. And since 1950, we've seen urban populations grow at a much faster rate than rural populations. The United Nations Department of Economic and Social Affairs World Urbanization Prospects report pre-dicts that we're about to see this continued urban growth com-plemented by a decline in rural populations. The end result is a planet of cities, surrounded by arable land, first in the developed and later in the developing world.[3]

It may not be obvious to people living in the developed world, but a city like Lagos—with all its obvious downsides—is an extremely appealing destination for Nigerians from rural areas.

In most developing-world cities, the schools and hospitals are far better than what's available in the rest of the country. And even with high rates of unemployment, the economic opportunities in cities vastly outpace what's available in rural areas.

There's a more basic reason for the appeal of cities—they are exciting. A city dweller has more options: where to go, what to do, what to see. It's easy to dismiss this idea—that people would move to cities to avoid boredom—as trivial. It's not. As Amartya Sen argued in his seminal book *Development as Freedom*, people don't just want to be less poor; they want more opportunities, more freedoms, more chances to better their lives. Cities promise options and opportunities, and they often deliver.

Harder to understand, in retrospect, is why anyone would have moved to London in the years 1500–1800, when it experienced rapid, continuous growth and became the greatest metropolis of the nineteenth century. The city had several major shortcomings, not least of which was an unfortunate tendency to burn down. The Great Fire of 1666, which left as many as 200,000 in the city homeless, was merely the largest of a series of "named fires" severe enough to distinguish themselves from the routine fires that threatened wood and thatch houses, packed closely together and heated with open coal or wood fires. It's likely that more Londoners would have been affected by the Great Fire but for the fact that 100,000—a fifth of the city's population—had died the preceding year of bubonic plague, which spread quickly through the rat-infested city.[4]

By the time of Dickens's London, the threat was less from fires than from the water system. Open sewers filled with household waste, as well as the manure of the thousands of horses used to pull buses and cabs, emptied directly into the Thames, the source of most of the city's drinking water. Cholera was common from the 1840s through the 1860s, and the smell of London during the hot summer of 1858 was so bad that it led to a series of parliamentary investigations. "The Great Stink," as historians know the event, finally led to the construction of London's sewer system in the 1860s.[5]

People flocked to cities in the eighteenth and nineteenth centuries, but not for their health. In the 1850s, the life expectancy for a man born in Liverpool was twenty-six years, as compared with fifty-seven years for a man in a rural market town.[6] But cities like London had a pull not unlike that of Lagos now. There were more economic opportunities, especially for the landless poor, and an array of jobs made possible by the international trade that flowed through the ports. For some, the increased intellectual opportunities provided by universities and coffeehouses was an attraction, while for others, the opportunity to court and marry outside of closed rural communities was the reason to relocate.

One reason to come to the city was to encounter the people you couldn't encounter in your rural, disconnected lifestyle: to trade with, to marry, to learn from, to worship with. You came to the city to become a citizen of the world, a cosmopolitan.

Just as Diogenes the Cynic went to Athens to debate the great minds of his day, cities have always attracted those in search of intellectual stimulus. If you wanted to encounter a set of ideas radically different from your own, your best bet in an era before telecommunications was to move to a city. Cities are powerful communication technologies, enabling real-time communication between different individuals and groups and the rapid diffusion of new ideas and practices to multiple communities. Even in an age of instantaneous digital communications, cities still enable constant contact with the unfamiliar, the strange, and the different.

Understanding the city as a communications technology helps make sense of the decisions Gibson and Stephenson made in using the city as a metaphor for cyberspace. Both were interested in the ways the Internet could bring the weird, dangerous, and unexpected (as well as the trivial, mundane, and safe) into a constant fight for your attention.

Gibson and Stephenson were both interested in virtual spaces, places where people were forced to interact, bumping into one another as they headed toward the same destinations. They believed that we would want to interact in cyberspace in some of the ways we do in cities, experiencing an overload of sensation, a compression in scale, a challenge of picking out signal and noise from information competing for our attention.

We hope that cities are serendipity engines. By putting a diverse set of people and things together in a confined place, we increase the chances that we're going to stumble onto the unexpected. Cities provide an infrastructure that should enable serendipity. From studying infrastructure and flow, we know that people rarely use infrastructures to their full capacity. Do cities really increase serendipity?

How Cities Work

In 1952, the French sociologist Paul-Henry Chombart de Lauwe asked a young political science student to keep a journal of her daily movements as part of his city study *Paris et l'agglomération parisienne* (Paris and greater Paris). He traced her paths onto a map of Paris and saw the emergence of a triangle, with vertices at her apartment, her university, and the home of her piano teacher. Her movements illustrate "the narrowness of the real Paris in which each individual lives."

That pattern of home, work, and hobby—whether it's a comparatively solitary activity like piano studies or the "great good place" of public interaction celebrated by the sociologist Ray Oldenburg—is a familiar one to social scientists. Most of us are fairly predictable. Nathan Eagle, who has worked with Sandy Pentland at MIT's Media Lab on the idea of "reality mining," digesting huge sets of data like mobile phone records, estimates that he can predict the location of "low-entropy individuals" with 90–95 percent accuracy on the basis of this type of data. (Those

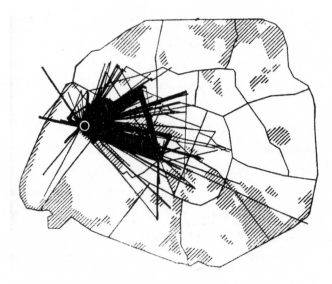

Walking routes of a young woman from the 16th
arrondissement over the course of a year.

of us with less predictable schedules and movements might be
only 60 percent predictable.)[7]

We might choose to see our predictability as evidence of con-
tentment and lives well lived. Or we can react as the situationist
cultural critic Guy Debord did and decry the "outrage at the fact
that anyone's life can be so pathetically limited."[8] One way or
another, the likelihood that we will be confronted with one of
these maps is increasing.

Zach Seward, outreach editor for the *Wall Street Journal*, is a
heavy user of Foursquare, the site that tracks your "check-ins" at
public places so it can recommend restaurants and destinations
to you. As he checks in at venues in and around New York City,
he generates a "heat map" of his wanderings. It's easy to see a
heavy concentration around Manhattanville, where he lives, and
Midtown, where he works. With a bit more work, we can see that
he enjoys hanging out in the East Village, rarely strays into the
"outer boroughs" except to fly from LaGuardia Airport and to
watch baseball games—the one venue he has checked into in the

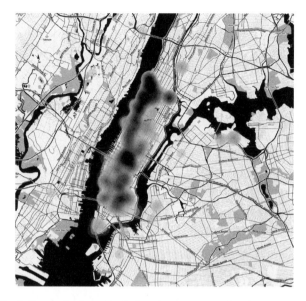

Zach Seward's New York City Foursquare check-ins, 2010.

Bronx is Yankee Stadium. And because he has visited one of New York's stadiums and not the other, we can bet he's a Yankees fan.[9]

If you're using Foursquare, you're broadcasting the data that can be used to make a map like this one. The Greek graduate student Yiannis Kakavas has developed a software package called "Creepy" designed to allow users—or people watching users—to build maps like this one from information posted on Twitter, Facebook, Flickr, and other geolocated services. Creepier, perhaps, is discovering that even if you don't use any of these services, you're leaking this type of data simply by using a mobile phone. While you may not be interested in suing your mobile phone provider to obtain this data, as the German politician Malte Spitz did, it probably has highly accurate data on your movements, which could be released to law enforcement on request or perhaps used to build a behavioral profile to target ads to you.

Seward took a close look at his Foursquare check-ins and discovered they provide information he hadn't realized: his race. He overlaid his check-ins in Harlem over a map that showed the racial composition of each block and discovered that "his" Harlem is almost exclusively blocks that are majority-white. As he observes, "Census data can describe the segregation of my block, but how about telling me how segregated my life is? Location data points in that direction."[10]

Seward is not a racist, nor is he "pathetically limited," as Debord suggests. We all filter the places we live into the places where we're regulars and the ones we avoid, the parts of town where we feel familiar and where we feel foreign. We do this based on where we live, where we work, and whom we like to spend time with. If we had enough data from enough New Yorkers, we could build maps of Dominican New York, Pakistani New York, Chinese New York, as well as black and white New Yorks. The patterns we trace through our cities show the effects of who we are, whom we know, and what we do; taken as a whole, they become maps of personal and group homophily.

When we talk about cities, we recognize that they're not always cosmopolitan melting pots. We acknowledge the ethnic character of neighborhoods, and we're conscious of ghettos that get separated, through a combination of physical structure and cumulative behavior, from the rest of the city. As a society, we hope for random encounters with a diverse citizenry to build a web of weak ties that increases our sense of involvement in the community, as Bob Putnam suggested in *Bowling Alone*. And we worry that we may instead hunker down when faced with a situation in which we feel like outsiders, as Putnam's more recent research suggests.

In our online world, isolation can occur as well, of course. As we saw in chapter 3, the biases of reported and curated media make some parts of the world less visible than others. Navigating through search brings our personal biases to the front. I get to learn about topics I care about—sumo, African politics, Vietnamese cooking—but I probably miss topics that I needed to know about because I was paying more attention to my interests and less to reporters and curators.

A recent wave of web tools tries to guide us to novel content by examining what our friends care about. Community-based tools like Reddit and Slashdot have formed communities around shared interests and direct us to stories the community agreed (through voting systems and reputation mechanisms) are interesting and worth sharing. Twitter and especially Facebook work on a much more personal level. They show us what our friends know and believe is important. Or as Brad DeLong puts it, Facebook offers a different answer to the question "What do I need to know?"—"You need to know what your friends and your friends of friends already know that you do not."[11]

Unless you've got a remarkably diverse and well-informed set of friends, there's a decent chance that their collective intelligence has some blind spots. The *Guardian* columnist Paul Carr tells an instructive story about returning to a San Francisco hotel room and being baffled that it, and the rest of the hotel, hadn't been cleaned that day. The hotel workers were protesting the Arizona immigration bill, SB1070, and while there was extensive conversation about the protests and the legislation on Twitter, they weren't taking place on feeds Carr followed on Twitter. By missing the protests until they manifested themselves as an unmade bed in his room, Carr realized that he was living in "my own little Twitter bubble of People Like Me: racially, politically, linguistically and socially."[12] Does that bubble provide us with the serendipity we hope for from the web? If not, we need to find ways to escape it.

Serendipity

Robert K. Merton devoted a book, written with his collaborator Elinor Barber and published posthumously, to the topic of serendipity. This may seem an odd exploration for a celebrated sociologist, but then again one of his many contributions to the field was an examination of "unintended consequences." These unintended consequences are often side effects of a successful intervention; for example, the introduction of rabbits to Australia provided a key food source for early white settlers, but inadvertently created a pest to farmers so severe that the Australian government was forced to build a 2,000-mile rabbit-proof fence to prevent crop destruction.[13]

Serendipity, at first glance, looks like the positive side of unintended consequences, the happy accident. But that's not what the term meant, at least originally. The word was coined by Horace Walpole, an eighteenth-century British aristocrat, fourth earl of Orford, novelist, architect, and gossip. He's remembered primarily for his letters, forty-eight volumes' worth, which offer a perspective on what the world looked like through the eyes of privilege.

In a letter written in 1754, Walpole tells his correspondent Horace Mann about an unexpected and helpful discovery he made, spurred along by his deep knowledge of heraldry. To explain the experience, he refers to a Persian fairy tale, *The Three Princes of Serendip*, in which the titular characters were "always making discoveries, by accidents and sagacity, of things they were not in quest of."[14] Walpole's neologism is a pat on the back—he's congratulating himself both for a clever discovery and for his sagacity, which permitted the discovery.

Useful as the concept is, the word "serendipity" didn't come into wide use until the past couple of decades. By 1958, Merton tells us, it had appeared in print only 135 times. In the next four decades, it appeared in book titles 57 times and graced newspapers 13,000 times in the 1990s alone. A Google search turns up 11 million pages (and counting) with the term, including restaurants, movies, and gift shops named Serendipity, but very few on unexpected discovery through sagacity.

Merton was one of the major promoters of the word, writing about "the serendipity pattern" in 1946 as a way to understand unexpected scientific discoveries. Sir Alexander Fleming's discovery of penicillin in 1928 was triggered by a spore of *Penicillium* fungus that contaminated a petri dish where he was growing *Staphylococcus* bacteria. While the mold spore landing in the dish was an accident, the discovery was serendipity. Had Fleming not been cultivating bacteria, he wouldn't have noticed a stray mold spore. And had he not had a deep understanding of bacterial development—sagacity—he might not have noticed the antibiotic properties of *Penicillium* and developed the most important advance in health technology of the first half of the twentieth century.

Louis Pasteur observed, "In the fields of observation chance favors only the prepared mind." Merton believed that serendipity emerged both from a prepared mind and from circumstances and structures conducive to discovery. In *The Travels and Adventures of Serendipity*, he and Barber explore discovery in a General Electric laboratory under the leadership of the chemist Willis Whitney, who encouraged a work environment that focused as much on fun as on discovery. A healthy blend of anarchy and structure was necessary for discovery, and overplanning was anathema, since "the policy of leaving nothing to chance is inherently doomed by failure."

The idea that serendipity is a product both of an open and prepared mind and of circumstances conducive to discovery can be traced back to the story referenced by Walpole in 1754. The three princes were deeply learned in "Morality, Politicks and all polite Lerning in general," but they did not make their unexpected discoveries until their father, the emperor Jafer, sent them out from his kingdom to "travel through all the World, to the end that they might learn the Manners and Customs of every nation." Once the well-prepared princes met circumstances conducive to discovery, unexpected and sagacious discoveries occurred: the identity of a royal poisoner, the strategy to defeat a mysterious giant hand that threatens a kingdom.

When we use the word "serendipity" now, it usually means "a

happy accident." The parts of the definition that focus on sagacity, preparation, and structure have slipped, at least in part, into obscurity. As the word has changed meaning, we have lost sight of the idea that we could prepare ourselves for serendipity, both personally and structurally. I suspect that we, and even Merton, understand those preparations poorly. And as my friend Wendy Seltzer, a legal scholar, pointed out to me, if we don't understand the structures of serendipity, it appears no more likely than random chance.

Designing for Serendipity

If we want to create online spaces to encourage serendipity, we might take some lessons from cities.

In the early 1960s, a fierce public battle erupted over the future of New York City. The proximate cause of the battle was the Lower Manhattan Expressway, a proposed ten-lane elevated highway that would have connected the Holland Tunnel (which links Manhattan and New Jersey under the Hudson River) to the Manhattan and Williamsburg Bridges (which cross the East River and connect Manhattan to Brooklyn). Plans for the highway required the demolition of fourteen blocks along Broome Street in Little Italy and SoHo and would have displaced roughly two thousand families and eight hundred businesses.

The proponent of the plan was Robert Moses, the legendary and influential urban planner responsible for much of New York's park and highway systems. His fiercest opponent was Jane Jacobs, activist, author, and chairperson in 1962 of the Joint Committee to Stop the Lower Manhattan Expressway. The lasting legacy of Jacobs's opposition to Moses is both the survival of Broome Street and her masterwork, *The Death and Life of Great American Cities*, a critique of "rationalist" urban planning and a manifesto for preserving and designing vibrant urban communities.

Jacobs framed many of her battles over urban planning by asking whether cities were for the benefit of cars or of people, suggesting that Moses was indifferent to the people he proposed to displace. A slightly less biased frame might submit that Moses took a bird's-eye, citywide view of urban planning while Jacobs offered a pedestrian's-eye, street-level view of the city. From Moses's point of view, one of the major challenges of New York City was allowing people to move rapidly from their homes in the suburbs to business districts in the center, and back out to the "necklace" of parks he'd painstakingly constructed in the outer boroughs.

In her critique of Moses, Jacobs offers two lines of questioning. One set of questions is political: Whom is a city for, and whose needs get considered when design decisions are made? By challenging Moses's role as an objective, disinterested expert, she invites readers to consider the unstated political biases in Moses's decision, for the wealthy suburbanites, against the poorer inner-city residents.

The other set of questions is about unintended consequences. For example, Moses's principle of separation of uses—residential neighborhoods separate from business districts, separated from recreation areas—may have the consequence of making cities less vital. What makes cities livable, creative, and ultimately safe is the street-level random encounters that Jacobs documented in her corner of Greenwich Village. In neighborhoods where blocks are small, pedestrians are welcome and there's a mixture of residential, commercial, and recreational destinations, there's a vibrancy that's absent from planned residential-only communities or from city centers that empty out when offices close. That vibrancy comes from the ongoing chance encounter between people using a neighborhood for different purposes.

Jacobs's vision of a livable city has had a major influence on urban design since the early 1980s, with the rise of "New Urbanism" and the walkable cities movement. These cities tend to favor public transit over private automobiles, and create spaces that encourage people's paths to intersect, in mixed-use neighborhoods and pedestrian-friendly shopping streets. As the urban planner David Walters observes, they're designed to help individuals linger and mix: "Casual encounters in shared spaces are the heart of community life, and if urban spaces are poorly designed, people will hurry through them as quickly as possible."[15]

If there's an overarching principle to street-level design, it's a pattern of designing to minimize isolation. Walkable cities make it harder for you to isolate yourself in your home or your car, and easier to interact in public spaces. In the process, they present residents with a trade-off—it's convenient to be able to park your car outside your home, but walkable cities are suspicious of too much convenience. The neighborhoods Jacobs celebrates are certainly not the most efficient in terms of an individual's ability to move quickly and independently. Vibrancy and efficiency may not be diametrically opposed, but the forces are clearly in tension.

Cities embody political decisions made by their designers, as well as the unintended consequences of those decisions. So do online spaces. These days, urban planners tend to be transparent about their agendas. They will declare an intention to create a walkable city with the logic that they believe increased use of public space will improve civic life. In the best of cases, planners test to see what works and report failures when they occur—the persistence of private car use in walking cities, for instance.

It's not always easy to get the architects of online tools to articulate the behaviors they're hoping to enable and the political assumptions that inform those decisions. Sometimes, the architects may not recognize the assumptions they are bringing to the table. Twitter began as a project management tool for distributed work groups and grew into a powerful network for sharing ideas and links. Because there's so much traffic on the network, efforts to archive and index tweets have been difficult to get off the ground.[16] While there's increasing evidence that Twitter can have political importance, the ephemeral nature of Twitter conversations means important events that unfold on social media aren't included in search engines and disappear over time. Was this an unintended consequence of Twitter's design, or a conscious effort to make online communication lighter, faster, and less permanent?

Other architects are clearer about their agendas, but can find their unintended consequences uncomfortable. Unlike other online networks that permitted users to log on using pseudonyms, Facebook has always required users to appear online using their real names. This policy dates from the site's origins as a replacement for paper "facebooks" issued at universities to help students meet one another. As Facebook became a popular tool for activism in repressive countries, human rights advocates warned that the policy could endanger the safety of dissidents.[17] Facebook has been inflexible, arguing that the company's "real name culture" is essential to maintaining a high quality of discussion on the site.

Online spaces need their Jane Jacobses to identify hidden politics and warn of unintended consequences. But they also need a better sense of historical context.

Some of the people who design online spaces are trying to increase exposure to a diverse range of information and to cultivate serendipity. But it's difficult to accomplish, in part because it's too easy to start from scratch in building online spaces. An urban planner who wants to make changes to a city's structure is hemmed in by a matrix of forces: a desire to preserve history, the needs and interests of businesses and residents in existing communities, the costs associated with executing new projects. Progress is slow, and as a result we have a rich history of cities we can study to see how earlier citizens, architects, and planners have solved these problems.

For those planning the future of Facebook, it's hard to study what has succeeded and failed for MySpace, in part because an exodus of users to Facebook has largely turned MySpace into a ghost town. It's harder yet to study earlier communities, like LamdaMOO, a text-based virtual world hosted on the servers of the legendary corporate research lab Xerox PARC. I often find myself nostalgic for Tripod, the proto-social network I helped build in the late 1990s. The admirable Internet Archive includes several dozen snapshots of pages on the site from 1997 to 2000, which gives a sense for the changing look and feel, but offers little insight into the content created by the 18 million users of the site in 1998. Tripod's more successful competitor, Geocities, disappeared from the web entirely in 2010—its legacy is less than 23,000 pages stored accessible through the Wayback Machine, which threw up its hands at the impossible task of archiving the vast site in mid-2001.

If we learn from real-world cities instead of abandoned digital ones, what lessons might we take?

Design for Encounter

In her celebration of street life, "the ballet of the good city sidewalk," Jacobs emphasizes the importance of using the same spaces for diverse purposes. Her neighborhood works because people pass through constantly. The random encounter with the shopkeeper as she leaves her house to head for work is possible only because he works where she lives.

Virtual environments like Facebook support many different modes of use. It came as a great surprise to most Americans, using Facebook to plan weekend activities with friends or to catch up on the developments in the life of an old flame, that Colombians were using Facebook to organize a protest against FARC rebels.[18] The Americans making weekend plans and the Colombians planning marches were in the same "space," but were utterly invisible to one another, unless they happened to be in one another's circles of friends. As a result, Facebook can feel like a single-purpose space, designed for whatever purpose you've chosen to put it to. You're part of a massive public space, but you're isolated by whom you know and what you do, much as early users of Facebook were largely isolated into campus-specific conversations.

Most American Facebook users didn't want to trip over highly political discussions on their way to tend their gardens in Farmville. But the Jacobs/Moses debate suggests we need to be cautious of architectures that offer convenience and charge isolation as a price of admission. It's more convenient for me to drive from my Cambridge apartment to my office, but the drive isolates me from the neighborhoods I pass through. As we examined in chapter 3, critiques of social networks like Eli Pariser's "filter bubble" argument worry about what we, individually and societally, lose from isolation. With Google's personalized search and Facebook's algorithmic curation of news from our friends, our online experience, he argues, is an increasingly isolated one, which threatens to deprive us from serendipitous encounter. Filter bubbles are comfortable, comforting, and convenient; they give us a great deal of control and insulate us from surprise. They're cars, not public transit or busy sidewalks.

With the rise of Facebook's "like" button on sites across the web, we're starting to see personalization come into play even on heavily curated sites like that of the *New York Times*. I can access whatever stories I want, but I also get signals regarding which of my friends have "liked" the story I'm reading, and what other stories they've also liked, as well as an algorithmic "recommended for you" list, generated for each subscriber. It's not hard to imagine a future where "like" informs even more information spaces. In the near future, I expect to be able to pull up an online map of whatever city I'm visiting and see my friend's favorite restaurants overlaid on top of it. I can already, using a service called Dopplr, which shares my movements and travel tips with friends, but I expect to see this functionality emerge as a default in Google Maps at some point soon.

Whether that scenario is exciting or troubling depends on whether I see only my friends' recommendations. If I also can see the favorites of other communities, that's a different story. As Pariser observes, the filters we really have to worry about are those that are opaque about their operations and that are on by default. A map of Vancouver that I can choose to overlay with my friends' recommendations is one thing; a map that recommends restaurants on the basis of paid advertisements and doesn't reveal this practice is another entirely. The map I want is the one that lets me shuffle not just through my friends' preferences but through annotations from different groups: first-time visitors to the city; longtime Vancouverites; foodies; visitors from Japan, Korea, or China.

If I'm visiting Vancouver, the most convenient map is the one that knows what I like and helps me find it. It knows that I'm a fan of ban minh sandwiches, locally brewed beer, and record stores with cheap, used vinyl. It's not hard to build this map today, and enhance it with knowledge of what my friends have enjoyed on their trips to the city. Before the plane lands, I can be presented with "My Vancouver." But unless I can look through that map and see the many different ways people are experiencing the city, I risk trading away the possibility of discovery for a comfortable form of isolation. As we design online spaces, we need to think through the dangers of making those spaces too comfortable, too easy, and too isolated.

Desire Lines

In any populated area, people create paths between the places where they are and the places where they want to be. These ad hoc paths, which urban planners call "desire lines," reflect a human tendency toward efficiency (or laziness). But they also provide valuable information about where people want to go and how

they want to get there. Smart designers have taken to photograph-
ing spaces after snowfall, or collecting time-lapse photographic
images so that they can lay paths atop the desire lines rather than
fight a losing battle between human behavior and lush grass.

Desire lines are a way that people inscribe themselves on places,
the accumulation of human behavior leaving visible traces. The
cupping of stairs as thousands ascend and descend, the marking
of a sidewalk with cigarette butts, discarded gum and dirt, and
the patina acquired through hands holding a rail are all subtle,
important signals about where people go and don't go, what they
do and don't do. As we walk through a city, evidence of human
activity bombards us: this park, crowded with strollers, is popular
with parents and children, while that one, littered with bottles,
attracts a different clientele. Whether the signals are generated in
real time, by the crowds outside the popular lunch joint, or over
time, they tell us stories about how people actually behave, not
just how designers and planners hoped they would behave.

As people began using Facebook to promote bands and brands,
as well as to keep up with high school friends, the company introduced a new kind of online space, distinct from the standard personal profile: a Facebook page. You could now become a fan of a
musician, a public figure, a movie, or another cultural phenomenon that had a page. Shortly after Pages were introduced in 2007,
Facebook offered an alphabetical directory of them, alongside
the little-used but fascinating alphabetical list of all Facebook
users, perhaps the world's largest virtual phone book.

Turn to a page of the directory, and you could see what's most
popular within a given letter. The letter *v* featured Vin Diesel
and Victoria's Secret, but also some more obscure celebrities: the
Turkish singer Volkan Konak, motorcycle racer Valentino Rossi,
Filipino talk show host Vice Ganda, and Mexican viral video
producers Vete A La Versh. The Pages directory was Facebook's
form of desire lines, revealing the topics most interesting to the
service's massively global user base, and a glimpse of the sorts of

figures who are celebrities in Nigeria, Colombia, or Vietnam, but not in North America or Europe.

Some months back, I went to Facebook to show the directory as part of a presentation about serendipity and discovery, and found something very different—a single page that featured a small number of pages that I'd expressed an interest in, and a selection of topics Facebook thought I might find interesting. Because I have many friends in the Middle East and because I conducted this experiment during the height of the Arab Spring, most of the pages recommended were Egyptian political organizations. I logged out of Facebook and visited the Pages directory not as Ethan Zuckerman but as a random user from an IP address in Massachusetts, and got a page filled with Boston sports teams and Dunkin' Donuts. Saving me the inconvenience of sorting the Red Sox from Real Madrid, Facebook buried its desire line information under a layer of forced customization, offering me an experience that's more comfortable, but less conducive to discovery.[19]

Not all social media services have taken the same approach. Twitter's trending topics, featured on the main page of that service, offer a glimpse into conversations that you would miss otherwise. You may not know what "Cala Boca Galvão" means when you see it as a trending topic, but it's an invitation to learn more. For some topics, Twitter has offered a one-line summary of the topic, making it easier to understand the significance of a topic in an unfamiliar language. It can also be an invitation to put up filters and learn less; Twitter offers you the ability to choose to see only trending topics from your local area, if you're uninterested in the Twitterings of people in Brazil or Japan. And trending topics favor fast-breaking news over events that take a long time to unfold, like the Occupy movement, making it easier to discover some trends than others.[20]

Simply revealing that a conversation is taking place does not guarantee that it will become any more inclusive. Martin Wattenberg and Fernanda Viegas, two of the world's leading information designers, tried a trending-topics experiment on a holiday weekend in 2010. They looked at the ten topics trending in the United States and then attempted to figure out who was involved in the conversations by looking at the profile pictures of people participating. They discovered a very sharp racial divide. Most trending topics were dominated by African American youth, who've been early adopters of the service and who often use Twitter quite differently from other users. The topic #inappropriatechurchsongs is filled with humorous suggestions for tunes your pastor won't be playing in church this weekend, because they're either racy puns on existing hymns or amusingly inappropriate suggestions. There's nothing preventing white Twitter users from joining the conversation, but Wattenberg and Viegas found that ninety-four participants in the thread were black, and six were white. Another topic—"oil spill"—skews differently, with several dozen white and three black authors. And "polyamorous" appears to be something Americans can all talk about, black or white.

Local Maxima

Twitter's trending topics are a form of local maxima, a sign of what's popular at a moment in time for a particular population. They're useful precisely because they're not global maxima. If Twitter showed us what was most popular across the whole service, we would encounter an endless stream of Justin Bieber memes. By looking for what's hot today, we encounter a mix of the expected and the surprising, the news we knew was breaking, and the news we're surprised to discover.

My friend David Arnold has emerged as a celebrity in the molecular gastronomy world as the director of culinary technologies for the French Culinary Institute in New York City. He's the guy who infuses gin with quinine in a whipped cream maker, carbonates it, and serves it in a liquid nitrogen–chilled glass to offer a gin and tonic that's entirely gin.[21] I knew him as a high school student, before his culinary fascinations became apparent, and when I knew him he was a big Bob Marley fan. Most high school Bob Marley fans are badly deluded white guys who have secret dreams of becoming dreadlocked Rastafarians. But Arnold just liked the music and had an apparently limitless collection of early Marley recordings.

I once asked Arnold how he'd started listening to Marley. "I went to the record store and realized there was this whole section—reggae—that I knew nothing about. I looked in, and this guy Marley had more albums than anyone else. I figured he was the best reggae guy, so I bought one of his albums and took it home."

The Dave Arnold algorithm is a remarkably effective way to explore an unfamiliar space where someone's done the work of classifying information for you. Don't know anything about French Impressionism? Try this guy Monet. If that doesn't work for you, you might give Renoir a go. If you're not a fan of either, there's a decent chance you can give up on the category and move on to Cubism or Abstract Expressionism. You may miss Delacroix, but you'll be able to sample some of the greatest hits of other artistic movements quickly, and then decide if you want to make a deeper dive. Computer scientists call this a "breadth first search," where you scan the horizon for promising matches before descending for a complete exploration of a category.

The local maxima strategy also works for wandering through a city, once you start thinking of places popular with different groups of people as categories to explore. Knowing that Times Square is the most popular tourist destination in New York may only be useful so you can be sure to avoid it. But knowing where most Haitian taxicab drivers go for goat soup is often useful data on where to find the best Haitian food. Don't know whether you like Haitian food? Try a couple of the local maxima—the most important places to the Haitian community—and you'll be able to answer that question pretty quickly. It's unlikely you'll dislike the food because it's badly prepared, since it's the favorite destination for that community.

There are limits to the local maxima algorithm—goat soup may not be the easiest Haitian cuisine to stomach. A bridge figure, perhaps a waitress who's experienced in introducing non-Haitians to her country's cuisine, might have a better suggestion. But following the cab driver is a first step in escaping your existing patterns of discovery. If you want to explore beyond the places your friends think are the most enjoyable, or those the general public thinks are enjoyable, you need to seek out curators who are sufficiently far from you in cultural terms and who've annotated their cities in their own ways.

Again, Twitter is a space where you can identify local maxima and let them curate your wanderings. Encounter an unfamiliar trending topic like "#M23"—the name of a rebel movement in Democratic Republic of Congo, allegedly supported by the Rwandan government—and you'll quickly discover that certain users are being amplified more than others. That's often because they're on-the-ground sources, or authoritative commentators on the topic in question. Someone like Laura Seay, an African studies scholar and professor at Morehouse University, wouldn't register on a top hundred or top thousand list of Twitter users. But if you follow a conversation about conflicts in Central Africa, she's likely to emerge as a prominent and widely quoted source. You might follow her to see whether you were interested in the topic, or to find the people she follows and quotes. Or you might quickly conclude that your interest in the space is limited and broaden your search for new inspirations.

Structured Wandering

The danger of the city is that we get trapped in Chombart de Lauwe's personally limited "real" Paris. The danger of online spaces is Pariser's filter bubble of comfortable media delivered by our friends. In each case, the solution comes in part from wandering, making a conscious decision to stray from our everyday path and experience something unfamiliar. It's possible to wander with a purpose. The flaneur strolls the streets as a strategy for encountering and understanding the city. We can wander in ways that seek serendipity. We take a familiar question into unfamiliar territory in the hope that, like Fleming, we will find that something unexpected and odd has landed in our petri dish.

Guy Debord, the French social theorist who decried our "pathetically limited" lives in the city, prescribed the "dérive," an unstructured drift through a landscape, as an intervention for overcoming these limitations:

> In a dérive one or more persons during a certain period drop their relations, their work and leisure activities, and all their other usual motives for movement and action, and let themselves be drawn by the attractions of the terrain and the encounters they find there. Chance is a less important factor in this activity than one might think: from a dérive point of view cities have psychogeographical contours, with constant currents, fixed points and vortexes that strongly discourage entry into or exit from certain zones.[22]

If you're worried about accurately sensing the psychogeographical contours of your city, or if you simply don't have an entire day to wander, Serendipitor might be for you. You tell Serendipitor, an iPhone and web application by Mark Shepard, where you are, where you need to be, and how long you have. Rather than calculating the shortest path between two points, as Google Maps usually does, Serendipitor prescribes a meandering path that will get you to your destination at the appointed time, but via a route that no human would ever rationally select. Some routes are ones you shouldn't take—a recent version of the application occasionally prompts you to flag down a passing car and see where it takes you! More an art piece than a practical application, Serendipitor is a useful provocation that if we've forgotten how to wander, we could always develop software to help us go astray.

Systems like Debord's and Shepard's are by their very nature arbitrary, and it's easy to dismiss them as random or silly. But sometimes you'll be forced to experiment with arbitrary systems because they're a feature of the local culture. In Ghana, virtually everyone you meet knows the day of the week he or she was born on. In Ashanti culture, and many other Ghanaian cultures, nicknames are based on the day of the week you were born. My son Drew, who arrived early on a Saturday morning, is Kwame. Had he been a bit quicker about coming into the world, he might have been a Friday-born Kofi. When I find myself in church in Accra, I know that I'll be called up to make my offering along with all the other Thursday-borns, the Yaws and Yaas. Each set of men and women will dance to the altar when their day is called and make their offerings; the pastor will then let us know which birth day offered the most, leading to good-natured ribbing about who is generous and who's stingy.

Being able to introduce myself as Yaw to a group of unfamiliar Ghanaians is a highly successful strategy for building arbitrary connections. In most groups, I'll meet at least one other Yaw, and we'll shake hands, make eye contact, exchange a few extra pleasantries. We're bound to find something to talk about—after all, we share a name. Whether or not a wise Ashanti chief consciously promoted the nickname system as a way of generating social capital, or whether Ghanaians simply kept the practice up because it has some ambiguous but positive social benefits, it works, it persists, and it introduces me to Ghanaians I would otherwise never know.

Lest you think social connection through something as arbitrary as the calendar is limited to tribes in West Africa, behold birth month groups. Pregnant women, often those carrying their first child, frequently join groups on the social networking site LiveJournal on the basis of their due dates. All the women scheduled to deliver in September join a group, and they're able to update each other and compare their progress and experience through the months leading up to delivery. Many of these groups persist long after the children are born as new mothers compare their children's development through infancy and beyond. The women involved have only two things in common—they're Live-Journal users, and they got pregnant at roughly the same time. Many groups include both atheists and people of faith, women from different races and ethnicities. The arbitrariness of birth month mixes people in an unusual way, and the deeply emotional experience of watching each other become a mother generates social capital and, often, unexpected friendships.

Arbitrary structures can be a useful mechanism for individuals to wander, not just for groups to discover new connections. Many years ago, Jonathan Gold set up a wholly arbitrary challenge for himself. Working as a copy editor for a legal newspaper, living on the famously multicultural Pico Avenue in Los Angeles, Gold decided that he would eat his way down the street, getting off the bus one stop earlier each week and taking himself to dinner at one of the Ethiopian, Korean, Cuban, Cambodian, or Jewish restaurants that line the street. His yearlong experiment turned into a brilliant article in the *LA Weekly*, titled "The Year I Ate Pico," and launched Gold on a storied career as a restaurant critic, one whose *Counter Intelligence* column focuses on low-end ethnic cuisine in LA. That work eventually led to a Pulitzer Prize, the first awarded to a food critic.

I see echoes of Gold's arbitrariness in a wonderfully strange project launched online by the *AllMetalResource* blog, a lead-ing weblog focused on heavy metal music. The authors behind the blog declared April 2011 "International Death Metal Month" and tried to find death metal bands from each of 195 UN-recognized nations. Their search didn't involve plane tickets, just lots of searches on YouTube. It turns out that death metal bands are a pretty sociable bunch, and are often looking for the feedback of fans in other countries. The organizers fell short of their goals, but they did make some remarkable discoveries, including the Botswanan band WRUST, part of a heavy metal scene whose members dress in black leathers with cowboy hats, looking like a postapocalyptic western movie set in southern Africa. Botswanan death metal may not be your thing, but that's really the point—pick a topic you're interested in and go global. The blogger Linda Monach makes dinner for her family, including her meat- and potatoes-eating father, each night. Bored with serving the same old food, she decided to spend a year serving burgers from around the world: an Albanian burger of lamb and flatbread, an Armenian creation that includes apricot preserves. The constraint: meat and bread in every meal. The possibilities? As wide as the span of human culinary creativity, or the selection at the local supermarket.[23]

Breaking the Metaphor

The difference between an online and a real-world walking tour reminds us that there's a danger in extending the geographic metaphor of the city too far. Compelling as we can make the narration of a walking tour or the promise of excellent goat soup, it still takes a long time to get from the Bronx to Staten Island. In digital spaces, on the other hand, we can change proximities. We can sort bits any way we want to, to reshuffle our cities any way we can imagine. We can create neighborhoods that are all waterfront, all park, all brick buildings, all eight-story buildings built in 1920 and discover who and what we encounter in these new spaces.

David Weinberger writes about the powers and potentials of the Internet. In his book *Everything Is Miscellaneous*, he explores what happens when information is freed from physical constraints. In the real world, a book can be shelved in only one hierarchy and one location at a time. But in a digital world, there's no limit to the ways you can shelve your collection or present your data. The analog world favors hierarchies and trees, while the digital world encourages us to "put each leaf on as many branches as possible."

Weinberger is one of the people behind the Harvard Library Innovation Lab, which, perhaps unsurprisingly, has been experimenting with reshuffling the library shelves, one of the most powerful structures we have to encourage constructive stumbling through an information landscape. Whether shelved by Library of Congress classification, or the older and quirkier Dewey decimal classification system, Weinberger, citing Clay Shirky, argues that the shelving of books in a library forces us to order knowledge in a single, universal way. This means we end up with biases that seem arbitrary and dated: Weinberger notes that 88 of 100 topics available in Dewey decimal coding for books about religion are reserved for Christian topics, while Buddhists and Hindus share a single catalog number.[24]

For all the flaws and peculiarities of this form of ordering knowledge, there are unanticipated positive consequences when we wander in an open-stack library. Because we expect books to be sorted by subject, we start with what we think we want to know and expand our search visually, broadening the topics we consider as our eyes move away from our initial search. As we scan the stacks, information about a book is available from its appearance—its age, its size. Width tells us whether the volume is brief or long, height is often a hint at whether a book contains pictures, as tall books tend to feature colored photos.

ShelfLife, a new tool developed at the Innovation Lab, offers the ability to virtually reshelve books according to these physical factors—size, width, height, age—as well as by data like subject, author, popularity with a group of professors or a group of students. The goal is to take what's useful about physical ways of organizing and the implicit information conveyed in those schemes and combine it with the flexibility of organizing digital information. Combining insights gained from studying the organization of cities with the ability to reshuffle and sort digitally may let us think about designing online spaces for serendipity in different and powerful ways.

How do we present the library shelf to maximize serendipity? How do we encourage readers to stray from the familiar toward the unexpected in a way that's likely to encourage discovery, not simply increase randomness?

Recommending the most popular books on the shelf simply becomes a form of social search. In the process, we inherit all the attendant problems: if others browsing the library are like us, we are trapped in another bubble of homophily. And even if a very diverse set of users uses ShelfLife, the system needs to know what we're interested in and what we already know about in order to make good recommendations.

Engineering serendipity requires a certain amount of surveillance.

Self-tracking and Self-discovery

Consider Dr. Seth Roberts, the man who tracks everything. Since 1980, he's been trying to cure himself of poor sleep, tracking his hours asleep, diet, weight, exercise, mood, and other factors. His experiments have tested a "mismatch theory," the idea that some of our discomfort with modern life stems from the ways in which our routines and practices vary from what humans were used to in the Stone Age. Roberts tried skipping breakfast (to mimic eating patterns of hunter-gatherers), watching human faces on TV in the mornings (to replicate the gossip and social contact that anthropologists believe characterized Stone Age mornings), and standing many hours a day. His meticulously documented find-

ings, correlating the quality of his sleep to his hours spent standing, persuaded him to move to a standing desk and do much of
his work while walking on a treadmill.

Eventually he discovered that he slept much better when he
stood for nine hours a day. By the time I met him, at a cocktail
party during the inaugural Quantified Self conference in Mountain View, California, he was experimenting with standing on one
leg until exhaustion multiple times a day. The bent-leg technique
was one he'd discovered by accident and was testing, trying random numbers of bent-leg stands a day to determine the optimum
number that correlated with restful sleep. At the moment, six was
looking like a pretty good number to him.[25]

Roberts represents two phenomena, taken to an extreme:
self-tracking and self-experimentation. Using tools like the Fitbit, which tracks each step you take, or Zeo, which tracks your
sleep states, self-trackers collect data about their bodies and their
moods and look for trends over time. Some self-trackers experiment on themselves, making changes in their diet, exercise, or
behavior, to see whether they sleep better or awake happier.

What would we learn from surveilling and experimenting with
ourselves in this way? We humans, for all our cognitive strengths,
are pretty poor at long-term self-awareness. We remember major
events more than minor ones, and much of what we do every day
blurs into the background. Tracking our behavior is a helpful
technique for shattering the illusions we all hold about ourselves.
I had been invited to the conference—hosted by one of the leaders of the Quantified Self movement, Gary Wolf—to talk about
some preliminary experiments I'd been conducting that looked at
a less frequently examined facet of the quantified self: consumption of media.

In my investigations of imaginary cosmopolitanism and the
Internet, I realized I needed data about what news stories people were seeing online and off-line, and what stories were capturing their interest. This information is fairly easy to obtain in

the broadest terms—hours of media consumed a day—but difficult to pin down in terms of specifics. Individual websites like the *Huffington Post* know what articles users read and how long they spend on a site, and advertisers are sometimes able to track users across multiple sites. But data on what individual news stories someone reads or what YouTube videos she watches is harder to aggregate. Tracking what people read or watch in analog media—radio or broadcast television, newspapers—relies on media diaries, logs that individuals keep of their viewing or reading behavior as well as set-top devices that track the programs played on a sample set of televisions. The data that emerge from tracking media consumption fuel a multibillion-dollar business. Even with access to pricey data sets from media-monitoring firms like Nielsen or Arbitron, it would be hard to answer the question "How much information from Africa did the average American get this week?"

Rather than paying a media analytics firm, I tried the Seth Roberts approach. For three months in the fall of 2010, I kept a diary of what I read, watched, and listened to off-line, and used a system called RescueTime to track my online behavior. RescueTime is designed as a productivity tool. It generates a scorecard judging how much productive time you spend at your computer versus "distracting" time—time spent writing versus watching YouTube videos, for instance. But you can use it in a less overtly judgmental fashion, simply looking at what captures your attention during the average day.

I found that my perception of myself and the actual person who populated my web browser history differ pretty sharply. I consider myself a globally focused guy: I chair the board of a Kenyan nonprofit organization, sit on boards of organizations focused on African journalism and global citizen media, and on many days I write about current events in different corners of the developing world. But that's hard to discern from my media consumption. What's more obvious when you review my online traces is that I've got a soft spot for Internet humor and that I spend an inordinate amount of time tracking my favorite football team, the Green Bay Packers. Globally focused news sites like those of the *New York Times*, the *Christian Science Monitor*, and South Africa's *Mail and Guardian* and my own site, Global Voices, received far less attention from me than reddit.com and ESPN. Comparing the amount of time I spent reading any news with the vast amount of time I spent reading and answering email was a soul-crushing discovery in and of itself.

Initially, I'd planned to blog about my weekly media consumption, but after the first week of tracking my behavior, I concluded that I was embarrassed even to share the files with my wife. If I wasn't seeking out international news online as often as I thought I was, I was stumbling over a surprising amount of it through a much older medium: radio. The most globally oriented days recorded in my media diary were ones when I spent a long time driving. National Public Radio's *Morning Edition* and *All Things Considered* feature heavy doses of international coverage, as does the BBC World Service, rebroadcast by many US public radio stations. The less control I had over what stories I encountered, the more international news I heard—and I noticed that many of my online searches for news started with stories I'd first encountered on radio.

Many self-trackers report that the act of logging their actions changes their behavior. If you're logging the food you eat during a day to track calories, the thought of recording the calories associated with a cheeseburger and fries can be sufficient to persuade you to order a salad instead. My experiences with tracking media were similar. I believe that encountering international news is important, and I was dismayed to see how little I sought it out. Pretty quickly, my visits to reddit and PackersNews.com decreased, and I found I was reaching for harder news sites during moments of cognitive downtime.

If a few million people track their sleep and their steps, fewer track their movements through a city as Zach Seward has. And fewer still track everything they read, hear, or encounter, if only because the challenges of collecting data are so great. But we would benefit greatly from tools that help us monitor what we see and help us understand what we know and don't know about.

Fitbit presents its wearers with a simple, stark blue number: the steps you've taken in a day. It's harder to lie to yourself that the walk around the block was the equivalent of a workout when Fitbit tells you it was only five hundred steps. Systems that show us where we go, whom we talk to, and what we read would give us data we can use to make changes. If we're living a life that, as Debord warns, is pathetically limited, we can make different choices and change our behavior.

But surveilling ourselves offers another possible benefit. A system that knows what you've seen can use that information to help you discover. Track what you're reading, and it becomes clear which local maxima have already been explored, and which could lead to unexpected discoveries. And perhaps we can make recommendations that are better than random, because our paths—through the city and through the Internet—reveal our desire lines, what we're searching for, as well as what we've not yet found.

This is not to suggest that engineering serendipity is as simple (or as complicated) as tracking what we encounter and seeking out information that's related, but unfamiliar. As we tackle those massive, complex problems, we have to consider another variable: our tolerance for risk.

Serendipity and Risk

We need serendipity because of our tendency to focus on the familiar, to miss what might be provocative and inspiring because it's unfamiliar and unknown. Implicit in the metaphor of wandering is the idea that serendipity is unpredictable, time-consuming, and far from guaranteed.

Much of the work that's been done on offering online recommendations focuses on reducing risk. We know that young social media users regularly rely on their network of friends for advice in making choices. If your friends collectively think the TGIFridays down the street is pretty good, it's a less risky choice than the unknown Turkish joint a block farther away. For high-cost choices, this aversion to risk is reasonable. Buying a car without knowing its provenance is almost certainly a bad idea, while buying a meal from an unfamiliar restaurant may well lead to an unexpected discovery.

Netflix, the online video rental company, has used recommendation systems to help customers discover movies they might not otherwise rent. This is a critical problem for the company, because many customers sign up, rent the few dozen movies they'd hoped to see in rapid succession, and then cancel their memberships, costing Netflix their monthly subscription fee. If Netflix can offer high-quality recommendations, it increases its odds of retaining a customer.

Netflix recommends movies on the basis of an idea called "collaborative filtering." In collaborative filtering, you express a set of preferences—a few movies you like, a few movies you dislike—and the system looks for other users who have tastes similar to yours. It collects a set of their favorite movies, then recommends ones you've not seen. The trick is calculating what users have tastes similar to yours.

One popular method for calculating this is a technique called "cosine similarity." A computer program collects your ratings of a set of movies and compares those preferences to all other sets of ratings. If your ratings are identical to another user's—that is, you both gave *Casablanca* five stars and *Mission: Impossible* none—you score a one. If you have no films in common, you score a zero. The actual math behind this is wonderfully cool, if slightly mind-bending. Imagine a world with only two movies— *Casablanca* and *Mission: Impossible*. I give *Casablanca* five stars

and *M:I* one. Put a point on a graph at (5, 1)—*Casablanca*'s our
x axis, *M:I* our *y* axis—and draw a line that passes through (0, 0)
and (5, 1)—that's the vector that represents my preferences.

Now pretend you liked *M:I* a lot and thought *Casablanca* was
overrated. You get a point at (1, 5), and a vector from (0, 0) to (1,5)
represents your preferences. The angle between your vector and
mine is a measure of how similar we are, and taking the cosine of
that angle provides a simple way to scale the value to be between
0 and 1 for angles between 0 and 90 degrees. The rub, of course, is
that there are more than two movies in the world. Cosine similarity
adds a new dimension to our graph for each new term. So when we
compare your tastes to mine to see whether we like the same movies, we're playing with vectors that exist in 100,000-dimensional
space, one for each of the movies in the Netflix collection. Don't
bother imagining 100,000-dimensional space; it will make your
head hurt. Just imagine 3-dimensional space and think about two
vectors that each emerge from 0, 0, 0 and each pass through an
arbitrary point in positive *x*, *y*, *z* space; it's easy enough to imagine measuring the angle between those two vectors. Then take it
on faith that, mathematically, you can do the same thing in many-
dimensional space.

Finding movies by using linear algebra can lead to some deeply
unexpected results. You like old Steve Martin movies and the
Japanese cartoon FLCL? Me too. And I like Wim Wenders's epic
Road Movies. There's not a sane system based on cinema history that would propose New German Cinema when prompted
with American slapstick and Japanese anime as inputs . . . but a
collaborative filtering system just might, if a few users with my
tastes are represented in the system.

Collaborative filtering works, and works well. If you've ever
purchased a book recommended by Amazon based on your
past purchases, you've benefited from the technique. But Netflix
wanted it to work better. So it offered a $100,000 prize to any
team that could improve its algorithms sufficiently. Because Net-

flix has massive amounts of data about what movies people have rated, it's able to test these algorithms against actual preferences: on the basis of fifty ratings of other movies from a user, what's she going to think about *The Breakfast Club*? Compare your predictions to her actual behavior, and you know how well or poorly you're doing. Offer measurably better predictions than Netflix's in-house algorithm, and the prize is yours.

The winners of the contest were a team of computer scientists from AT&T and Yahoo! There was no single conceptual breakthrough that put them over the top. Instead, they found hundreds of small improvements they could make to the Netflix algorithms that, taken collectively, were a major improvement. Netflix paid them, the system improved, but we didn't learn anything particularly paradigm-shattering about collaborative filtering.

My friend Nathan Kurz wasn't part of the winning team, though early in the challenge his algorithms were placing in the top twenty on the leaderboard. Roughly halfway through the contest, Kurz realized that he and Netflix didn't see the problem of recommendation the same way. To win the Netflix challenge, your predictions of a user's ratings had to match its ratings quite closely. Knowing that a user would think a movie was worth three stars out of five—a "meh," in qualitative terms—is as important as knowing that a user would be likely to give a movie five stars. Netflix cares about this because it wants to predict what you're going to think about any film. For Kurz, that's a silly question: "Who rents a movie they're likely to give three stars? I want the movie that's going to change my life. I want to watch a movie that no one has ever heard of and fall so deeply in love that I want to meet the other people who've given it five stars because they're likely to be my soulmates."

As competitors in the Netflix prize discussed their progress, it became clear that a small subset of movies was extremely difficult for algorithms to handle. One of these movies is *Napoleon Dynamite*, a cult film about growing up in small-town America that

Netflix users appear to love or hate. Very few people give the film three stars—if you bothered to rate the film, you likely gave it a one or a five. Since winning the Netflix challenge involves predicting a user's preferences with a high degree of accuracy, solving the *Napoleon Dynamite* problem became central for many contestants. Most built systems that tried to minimize the influence of polarizing movies on other predictions—knowing that someone gave *Napoleon Dynamite* a five doesn't appear to help predict her taste for other movies.

For Kurz, movies like *Napoleon Dynamite* were a clue that there are other ways to offer collaborative predictions. You could swing for the fences, offering only films that might be fives for you, taking the risk that some—perhaps many—would fall short of the mark. As a user of Kurz's system, you would lead a life that was a lot less predictable, but potentially more rewarding and exciting.

Unfortunately, Kurz's life hit a patch of unpredictability before the close of the Netflix challenge. He caught the West Nile virus from mosquitoes living in a drainage ditch behind his home in Las Cruces, New Mexico, and spent two years recovering from the disease. While fighting the disease, Kurz experienced decreased cognitive capacity. "I couldn't write code," he says, "and I certainly couldn't design novel algorithms. I didn't know if I was going to get better, so I had to start thinking about another way to make a living."

Kurz chose cooking, and began making sorbet, using the sorts of highly technical processes only a Dave Arnold could love. His sorbets have no emulsifiers or binders. They're fruit juice, sometimes sweetened with beet sugar, frozen into dense, cold cylinders and scraped into sorbet by a machine called a Pacojet, which uses a fast-moving titanium blade to shave the ice into a two-micron-thick layer. The result is something that feels profoundly creamy while tasting of nothing but the primary ingredient.

True to his theories on recommendation, Kurz's sorbets

can polarize. Flavors like Almond Pink Peppercorn, Anaheim Chile, Sugar Snap Pea, Fennel Citrus, Nectarine Habañero, Rhubarb Ginger, and Coconut Thai Basil are not guaranteed crowd-pleasers, but in any sufficiently large group of people, you're virtually guaranteed to find someone who thinks that Beet Lemon is the best thing he's ever tasted.

"It works because we encourage people to taste all the flavors," Kurz tells me, as I finish a cup of Lemon Shiso in Scream Sorbet, his Oakland shop. "They're going to hate some of them, but usually there's something in there that they love. I don't ever want anyone to eat anything mediocre. I want them to taste their new favorite thing."

You can't discover penicillin unless some random mold spores drop into your petri dish. And you may need to tolerate a bite or two of Roasted Daikon (which Kurz considers his least-successful experiment) before discovering your love of Almond Pink Peppercorn (my personal favorite). To experience serendipity, we need to take the risks of failure, of frustration, of wasted time. If we're reengineering the media we encounter and the media we produce to encourage serendipitous discovery, the key may be to increase our tolerance for risk and to make it less painful—or at least more tasty—to fail.

In the next ten years, I expect that tools that enable serendipity, that help us stumble on unexpected and helpful information, will become as important a utility as search engines and social networks are today. At MIT, my students and I are working on systems that closely watch what you read online, and what content you choose to share, not to find users who are similar to you, but to help you find communities you know little about. We're exploring ways to find hundreds of communities in networks like Twitter or Facebook, and surface stories that prove interesting to otherwise unconnected groups. In other words, we're looking for local maxima that you might not otherwise discover.

There's vast work to be done in this space, whether it involves building tools that help readers and researchers see what they're seeing and missing or helping curators lead people into unfamiliar neighborhoods and weird parts of the Internet. Needed are both technological breakthroughs and new ways of approaching the problems of exploration and discovery. Better systems to visualize what we have already encountered are crucial, but so are tools that help us find translators and bridge figures who can contextualize what we're finding. Our first steps to designing for serendipity start with realizing that the ability to make novel connections is a new form of power.

THE WIDER WORLD

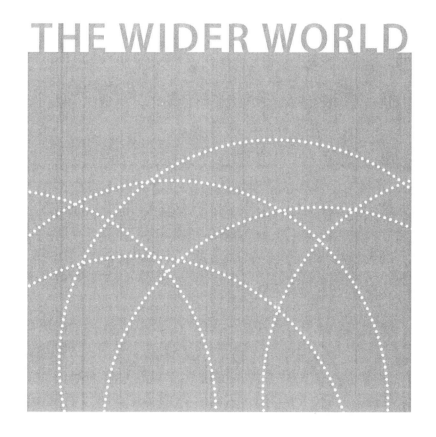

Chapter 8

THE CONNECTED SHALL INHERIT

THE GUITARIST NEAL SCHOEN HAD A PROBLEM. WHILE THE glory days of his band Journey ended in the mid-1980s, their anthemic arena rock ballads still attracted fans from around the world. In 2007, when Journey's hit "Don't Stop Believing" was used in the closing scene of a television drama,[1] *The Sopranos*, millions of American fans recalled youths spent in acid-washed jeans and wondered when Journey might tour again.

Schoen and his bandmates wanted to tour, but at the time Journey lacked a lead singer. After Steve Perry left the band in 1986, Journey had cycled through a long lineup of vocalists. Singing for Journey is no easy task: the band's hits, recorded with Perry at the helm, abound with technically demanding vocal lines sung by a powerful countertenor. For a dozen years—the post-Perry era—Journey fans had suffered through vocalists not quite up to the task. When he began his latest search for a new lead singer, Schoen wasn't looking for someone who would reinvent the band in his own image. He needed a highly talented, technical artist who could help Journey sound the way it had in its glory days.

Schoen turned to YouTube and began watching videos of 1980s cover bands, looking for the right vocalist.[2] Two days into his search, he stumbled onto an extraordinary performance. In the Hard Rock Café in Makati, one of the Philippine cities that make up the sprawling metropolis of Manila, a band called the Zoo was performing songs from the seventies and eighties soft rock canon: Air Supply, Night Ranger, and, of course, Journey. The guitarist was out of tune, the keyboard player stumbled over easy riffs, but the vocalist sounded exactly like Steve Perry. Neal Schoen had found Arnel Pineda.

Pineda was a city boy, born and raised in Sampaloc. His childhood had the features of a ballad a Journey audience might wave its lighters to. Pineda's mother had encouraged her son to compete in singing competitions, but when she died of heart disease when he was thirteen, Pineda found himself homeless, making his living collecting and selling bottles and scrap metal. His destitution didn't stop him from performing; at fifteen he was frontman for a local band, and by the time he was twenty-one his bands were winning contests in the Philippines. When Pineda caught Schoen's attention, he'd been performing in nightclubs for two decades. Although he had cut a couple of albums for the Filipino arms of international record companies, his name wasn't widely known outside of South Asian nightclubs and karaoke bars.[3]

Schoen had no way of knowing that the Philippines was the perfect place to look for his new lead vocalist. The long and complicated relationship between the United States and the Philippines ensures that American pop culture enjoys widespread exposure in Manila, and that most Filipinos speak English. In addition, a style of singing called *plakado*—the Tagalog word for "platter" or "record"—has been popular there since it was developed in the 1960s. Filipino singers reproduce recordings as faithfully as possible, and the highest form of praise for a live vocalist is that he sounded *plakado*, exactly as it did on the recording. *Plakado* reached even greater stylistic heights as electronic karaoke machines that scored vocalists on their precision became commonplace in Southeast Asia. All of that served to make the clubs and bars of Manila, in essence, a system engineered to produce vocalists who sounded exactly like Paul McCartney or Steve Perry.[4]

The videos of the Zoo performing in Makati had been uploaded by Noel Gomez, a friend and longtime fan of Pineda's, and Schoen reached out to Gomez to schedule an audition with Pineda. Gomez first had to convince Pineda that the invitation wasn't an elaborate prank, a process that ultimately involved placing a call to Schoen in California.[5] Then Pineda had to persuade the US embassy to grant him a visa to travel to his audition.

He got the visa, the audition, and the job. On February 21, 2008, Arnel Pineda debuted as the lead singer of Journey at the Viña del Mar International Song Festival in Chile, a concert broadcast live to 25 million television viewers. Pineda was an instant hit with Journey fans, who made the first album Pineda recorded with the band a platinum-selling success. What happened next was more unexpected. Not content with reviving Journey, Pineda has connected the band with a new audience—millions of Filipinos at home and abroad who love arena rock. The crowds at Journey shows now feature a mix of aging rock fans remembering their suburban youth alongside young Filipino Americans reveling in the success of their countryman. Journey's last concert video, released through Walmart, features the band playing for its most passionate fans. It's titled *Live in Manila*.

In a world where the Filipino lead singer of an American rock band wows crowds in Chile, it's the connected who shall inherit.

In our connected age, people who are able to bridge between cultures like Pineda have certain superpowers. They are able to sample from what's best in a variety of global cultures and recombine those influences in creative and novel ways. They can translate what's wonderful about a culturally specific art form and make it accessible to a new audience. There's no doubt that Pineda is supremely talented. But part of his genius comes from his background and role as a bridge figure.

The Changing Face of the CEO

The world's biggest corporations are seeking cosmopolitan leaders to build world-straddling businesses. Their reasons for selecting CEOs with a global perspective help us understand the power of diverse points of view and what we, individually and collectively, might gain from seeking cognitive variety through global awareness.

Multinational companies are becoming global companies, no longer rooted in one country and selling to the world, but seeking the best talent and leadership from wherever they work. We can see the transformation at work with two brands that symbolize American culture to much of the world: Coke and Pepsi. These American institutions are led by a Muslim from Turkey and a Hindu from India, respectively.

In 1980, Indra Nooyi was looking for work. After earning a BS in her native Madras, and a management degree from the Indian Institute of Management in Calcutta, she came to Yale to earn a second management degree. Living in the United States for the first time, she supported herself working an overnight shift as a receptionist at a campus building. When she landed an interview for a summer internship with the consulting firm Booz Allen Hamilton, she faced a dilemma: she could not afford a business suit. Rather than borrow one, she wore a sari, and she got hired.[6]

After stints at Boston Consulting Group and Motorola, Nooyi found herself at Pepsico in 1994. The global beverage giant faced two challenges—breaking into new markets in the developing world and diversifying the company's base of sales from snack foods to a broader and more healthful line of foods. Who better to lead the transformation than a vegetarian from Madras? Nooyi was named to the board of directors of Pepsico in 2001 and named CEO in 2006. While she is not without critics, Nooyi is widely recognized as one of the most powerful and successful executives on a global stage, and her name is floated as a candidate for positions as varied as the chief executive of the Indian manufacturing giant Tata and as a possible future head of the World Bank.[7]

While the Cola Wars don't rage as fiercely as they did in the 1980s, it's fair to say that Coca-Cola keeps a close eye on smaller rivals like Pepsi. And in 2008, Coke appointed Muktar Kent its CEO. Like Nooyi, Kent had lived, worked, and gone to school around the world. The son of a Turkish diplomat, Kent was born in New York, raised in the Asian countries his father was posted in, and educated in Turkey and the UK. His first positions with Coke took him to the United States, Rome, and Amsterdam, before he took on logistics for the company in Turkey and later in Central Asia. After twenty years at Coke, he left to head Efes, Turkey's largest beverage company, and returned in 2005, to head Coke's international operations.[8]

Coke and Pepsi may be ahead of the curve on CEO recruiting. A 2009 study by the management consultants Herman Vantrappen and Petter Kilefors calculates that 14 percent of Fortune 500 companies are headed by non-native CEOs, leaders from a country other than where the company is headquartered. On the one hand, that percentage seems small, given that in many large companies, the majority of employees work internationally, not where the company is headquartered. But the number of non-native CEOs was near zero a few decades past, which means the few dozen companies like Coke and Pepsi that have hired transnational executives like Nooyi and Kent are part of an emerging phenomenon.[9]

Companies may be looking to transnational CEOs in part because big companies are growing, expanding in the size of their operations and the markets they serve. When *Fortune* started tracking the world's largest companies in 1955, the top 500 firms had collective revenues comparable to 39 percent of US gross domestic product (GDP). While the Fortune 500 were big, they were less powerful, collectively, than the thousands of small, family-run firms that dominated the US and European economies. Not any more. The combined revenues of the Fortune 500 would now make them the second-largest economy in the world, and compare to 73 percent of the US GDP.[10]

Much of the growth comes from US, European, and Japanese brands finding new markets in emerging nations. As they have grown, these firms have become unfathomably complex. At Coke, Kent is responsible for over 140,000 employees in 206 countries and territories—and that represents only a fraction of the company's business; Coke relies on thousands of national and regional bottlers to get its product to consumers.

More than ever, the CEOs of the world's biggest companies have to manage and motivate an employee base that speaks a panoply of languages and competes in radically different markets. Finding ways to succeed frequently means putting aside long-standing assumptions drawn from the company's experience in established markets. This becomes all the more true when a company begins to falter.

Some firms have turned to transnational CEOs in times of trouble. The two leading Japanese firms led by non-native CEOs, Sony and Nissan, both brought new leaders in during a crisis. Carlos Ghosn, born in Brazil, of Lebanese descent, and educated in Paris, was working for Renault when the company took a major stake in Nissan in 1999. The Japanese auto manufacturer was deep in debt and making money on only three of its forty-eight models. Renault asked Ghosn to become CEO of Nissan in 2001, and he accepted, promising to return it to profitability in a single year.

He did so by shutting down plants, laying off 14 percent of the workforce, and ending the *keiretsu* system, in which Nissan owned substantial stakes in most of its parts suppliers. In the process, he violated virtually every established practice in the Japanese auto industry, firing workers who had expected lifetime employment and switching the corporation's working language from Japanese to English. Despite his audacity, many Japanese businessmen revere him. A 160-page comic book celebrates his success at Nissan, and a popular Japanese restaurant offers a Ghosn bento box, with rice and sushi shaped to resemble his face. Renault celebrated him by appointing him its CEO, while asking him to continue running Nissan. The twin jobs make Ghosn an extreme commuter, splitting his time between Paris and Tokyo.[11]

Nissan needed an outsider who could defy convention, make decisions that were consistent with the global best practices of the auto industry, even if they were unexpected within Japan. And Pepsi may have turned to Nooyi in the hopes that her knowledge of India would help the company navigate emerging markets. But the changing face of business leadership also suggests that hiring transnational CEOs may be about leveraging diversity.

One indicator is the rise of Indian CEOs on a global stage. Thirteen companies in the Fortune 500 are headed by executives born in India, which means India has produced more current CEOs than any country but the United States. And while Indian executives lead giant Indian firms like Tata and Mital, they also lead non-Indian giants like MasterCard, Citigroup, and Unilever. There are lots of reasons Indians are well positioned to lead multinational firms: they are fluent in English, the global language of business, and they cut their teeth in India's competitive and challenging home markets, with a government that deserves its reputation for creating red tape. But the most important reason Indians succeed in business may be the nation's profound religious, linguistic, and cultural diversity.

Ajay Banda, who heads MasterCard, and his brother Vindi, who headed Hindustan Lever from 2000 to 2005, credit their business success to their itinerant childhood. Their father, a lieutenant general in the Indian armed forces, moved to different parts of the nation every few years. "You had to adapt to new friends, new places," Vindi has said. "You had to create your ecosystem wherever you went."[12] Even Indians who didn't move around as much as the Banda brothers learned to live in a nation where neighbors spoke one of more than four hundred languages and practiced faiths as diverse as Jainism and Sikhism, where it was unrealistic to assume that your perspective on the world was shared by everyone around you.[13]

In the United States, the Partnership for a New American Economy, a bipartisan coalition of American business and political leaders focused on reforming America's immigration laws, has discovered that 18 percent of Fortune 500 businesses have at least one immigrant founder. More than 40 percent have founders who are immigrants or children of immigrants. These companies include Google, Intel, eBay, and Yahoo!, some of the pillars of the digital economy.[14] Why are new Americans so successful in building global-scale businesses?

America's immigration policies must have an influence, and it's possible they are sufficiently onerous that only the truly talented, hard driven, and best educated are given an opportunity to come to the United States. Or that immigrant parents push their children toward fiscal success harder than nonimmigrant parents. We might also consider that immigrants and their children are often deeply versed in the art of cultural bridging. The same skills that let an immigrant search for a job in an unfamiliar language, or allow a child to act as a cultural translator for her immigrant parents, can also give an executive an edge in building diverse teams or creating a corporate culture that values competing points of view. A bridge figure is more likely to be able to translate insights from other markets and less likely to be trapped listening only to the company line.

It's easy to dismiss a commitment to diversity as a form of tokenism, signaling that Pepsi takes the developing world seriously enough to hire an Indian executive, or as a form of political correctness. But hiring a CEO like Indra Nooyi also demonstrates, at the highest level of an organization, a commitment to organizational diversity. Given the massive investments companies are making to train and promote executives from around the world, and the responsibility they've put in the hands of Nooyi or Ghosn, we should consider the central importance some companies are giving to cognitive diversity.

Cognitive Diversity

When Ronald Burt explored the spread of good ideas at Raytheon (see chapter 6), he was focused on the creativity of individuals, concluding that the company's strongest thinkers were often bridges in an organizational structure, linking units in a corporation that might not communicate with one another. This structural position allows bridge figures to draw on a set of unfamiliar ideas and engage in creative thinking by adapting them for other uses. It's a powerful frame for addressing how individuals might help an organization solve complicated problems, but it leaves an important question unanswered: How should groups leverage diversity?

Scott Page, a scholar of complex systems at the University of Michigan, has spent much of his career examining this problem. In *The Difference*, a book written to share his complex mathematical ideas with a lay audience, Page offers some provocative insights. All things being equal, Page argues, a diverse team will solve problems better than an equally talented team of likeminded individuals. Furthermore, a diverse team selected at random will often perform better than a team composed of members who are individually highly skilled at solving a problem. In some circumstances, diversity trumps ability.

Page's argument doesn't extrapolate from experiences in the business or political world. Instead, he and his colleague Lu Hong set up a computer model where competing software programs called "agents" search for the best solutions (in this case, high numbers distributed at random, in a "map" of locations the agents can explore). Following individual, custom sets of rules, an agent would search a huge set of random numbers and locate those with the highest values. A team of agents would work together, sharing their findings, and would submit the highest number found in their collective explorations. Page and Hong then conducted a competition between teams, one team composed of the twenty agents who'd performed the best individually, and another composed of twenty agents chosen at random. The teams always performed better than a single agent did, but, surprisingly, the team of twenty randomly chosen agents consistently outperformed the team composed of high achievers.[15]

Randomly chosen teams win because of the local maximum problem. There are many possible good solutions to Hong and Page's problem, high numbers distributed in the set of random numbers. Their agents find the highest number they can in a limited period of time—the local maximum—then communicate with their teammates, so everyone can then turn in the biggest local maximum the twenty agents have found. This method works well as long as agents are searching different corners of the map. But if they're all searching the same space, they converge on a small set of local maxima, and miss other parts of the map.

Consider a real-world analogy. If I and a team of twenty people were asked to plant a flag on the top of the highest mountain we could reach within an hour's travel, I'd know what to do. I live near Mt. Greylock, the tallest mountain in Massachusetts, and I'd confidently plant my flag 1,063 meters above sea level. If my teammates all live in western Massachusetts, they'd end up on Greylock with me, because it's the tallest mountain for miles around, the local maximum. If one of my teammates lived in northern New Hampshire, we'd get very different results. She'd make for the top of Mt. Washington and plant her flag at 1,917 meters. And our team would do better still if one of our members lived in the Rockies, the Alps, or the Himalayas. If it's a competition between twenty rally drivers and expert mountain climbers who are all in the same, randomly chosen location, and twenty random individuals scattered around the world, the diverse team is going to win most of the time, especially if the mountain climbers are stuck in Nebraska.

Page and Hong's experiment relies on two important assumptions. One is that the problem is too hard for a single individual to solve optimally most of the time. For problems where a skilled individual can solve the problem herself most of the time, team diversity doesn't matter. Second, Page and Hong assume that people highly skilled at solving a problem tend to solve problems the same way. In their experiment, the agents that performed the best individually used very similar algorithms. But because they use similar techniques, they're clustered in one corner of the map of possible solutions, and they can be outperformed by a team of less-effective agents that can see more of the map.

The phenomenon is not unlike the "best and the brightest" problem. President John F. Kennedy was noted for surrounding himself with smart, young foreign policy advisers, whom the journalist and historian David Halberstam termed "the best and the brightest," in his book on the origins of the Vietnam War. These individuals were unquestionably bright, but they came from similar backgrounds—the best preparatory schools and elite universities, the same university and foreign policy jobs. Personality differences aside, on the whole they tended to solve problems in similar ways. Furthermore, they liked the solutions their fellow advisers proposed, because those solutions were comfortable, familiar, and likely to preserve harmony within the group. As a result, they didn't look for other perspectives or solutions, and reinforced each other's limited thinking, a process the psychologist Irving Janus termed "groupthink."[16]

A shared set of assumptions about China, the Soviet Union, and Vietnam proved disastrously wrong. Had the best and the brightest included a more cognitively diverse set of people, the mean intelligence of the group might have gone down, but their collective ability to solve a problem would have increased. Page tells us, "Even if we were to accept the claim that IQ tests, Scholastic Aptitude Tests scores, and college grades predict individual problem-solving ability, they may not be as important in determining a person's potential contribution as a problem solver as would be measures of how differently that person thinks."[17] Adding more Harvard graduates wouldn't have made Kennedy's team of advisers any smarter. Adding someone who could have challenged their thinking about China or Vietnam might have.

What kind of problems benefit from the cognitive diversity Page prescribes? Really hard ones. Page singles out the breaking of the German Enigma and Lorenz ciphers during World War II by a team at Bletchley Park in England. Legendarily eclectic, the team included not just mathematicians but also crossword puzzle experts, linguists, classicists, and ancient historians, and it drew on the intellectual resources of several Allied nations, including the remains of the vast British empire. It was also legendarily successful, consistently decrypting German messages and providing the Allies with military intelligence that Churchill so valued that he saluted the Bletchley Park staff as "golden geese that never cackled."[18] In the more recent past, Page suggests, the team that won the Netflix prize—seven computer scientists, statisticians, and engineers representing United States, Austria, Canada, and Israel—benefited from being a cognitively diverse group.[19]

Diversity and Dissent

Should your company hire an Austrian statistician or a Polish classicist? Page is careful to distinguish between identity diversity and cognitive diversity. Identity diversity refers to differences in gender, ethnicity, national origin, religion, language, and dozens of other factors. Cognitive diversity stems from differences in perspective and heuristics, the tools we use to solve problems. People from different backgrounds do tend to bring different perspectives and heuristics to the table, but there's not a one-to-one mapping between identity and cognitive diversity. An African American executive from rural Alabama and an Indian executive from Madras may think very similarly if they went to the same high schools, colleges, and executive training programs.

Hiring someone very different from the rest of your team in an effort to increase cognitive diversity also invites conflict. Page and Hong warn that "identity-diverse groups often have more conflict, more problems with communication and less respect and trust among members."[20] Warren Watson studied diverse teams in his management classes at the University of North Texas, placing students into work groups that were either all white American or a diverse pool where teams featured a white American, an African American, a Hispanic student, and a non-American student. He studied how well they worked together over the course of a semester, checking in at regular intervals. The diverse teams had more "process problems" than homogenous teams, conflicts over how the group should work together, and they were less successful in the short run. Over the course of the full semester, diverse and homogenous teams performed equally well, and diverse teams did a better job in finding a broad range of innovative solutions to problems.[21]

Even when a diverse team solves a problem better than other teams, its members may not enjoy the process as much. Katherine Phillips, who teaches at Northwestern University's business school, conducted a clever experiment in which teams of three sorority or fraternity members worked together to solve a murder mystery, combining pieces of information each had individual access to into a complete whole. Five minutes into the twenty-minute discussion, a new participant entered. In half the cases, it was a member of the same fraternity or sorority; in the other half, the new participant came from a different group. The groups where an outsider joined were significantly more successful in solving the mystery, but they were less confident than the homogenous team that they'd come to the right conclusion. Members of teams with an outsider also enjoyed participation in the exercise less than members of homogenous teams. Diversity made the teams more successful, but less comfortable for all those who participated.[22]

Phillips saw a significant effect from a minor difference—her participants were affiliated with different Greek organizations, but were all students attending the same university. The identity diversity we find in a global company or an urban neighborhood is likely to be much higher. Similarly, the discomfort we feel working together, whether on a new-product rollout or on a new community park, is higher as well.

Diversity and the Connected World

Some individuals who've grown up in multiple cultures learn to leverage their diversity of perspectives and heuristics into creative success in business or the arts, but this is far from guaranteed. And it's not clear how those of us who live and work near the place in which we grew up—the vast majority of people—can acquire cognitive diversity, or how we might embrace the benefits of diversity while minimizing the conflict and discomfort that so often accompany it.

It's a long road, with some possible shortcuts.

States, corporations, and individuals are all looking to the power of diverse perspectives as a path toward inspiration and toward tackling our most complex and pressing problems. And while the problems inherent in retooling a city or a country to benefit from diversity, while minimizing the stresses created by conflicting goals, are far from solved, we might look to cosmopolitan nations in the past and the present for inspiration.

At the turn of the seventeenth century, the Netherlands was the center of the modern world. The Dutch East India Company, the world's first multinational, was funded by the Bank of Amsterdam, the world's first central bank. Trade from as far away as Japan passed through the port of Amsterdam, and the city featured the first full-time stock exchange. The Dutch East India Company was far from an enlightened cosmopolitan actor in Asia; it used military force to negotiate favorable trading terms with the people it colonized. But Holland itself developed a reputation for connection, tolerance, and success.

The Dutch golden age was in part an accident of history. Spain's invasion of the Low Countries during the eighty years' war drove wealthy merchants and weavers from Antwerp to Amsterdam. Geography helped, too. Trading routes crossed through the state and brought grain from the Rhine to the Mediterranean and wine from France and Portugal. But the largest factor in the Netherlands' success may have been religious tolerance. As Catholic kings in Spain and France cracked down on nonbelievers in their lands, Protestants, Anabaptists, and Jews flocked to the Netherlands, bringing wealth, skill, and intellect. By allowing religious minorities to own land and property and to worship freely, the Netherlands attracted heretics from around the world. Some of them built great commercial empires, others academic institutions like the University of Leiden.[23]

In our day, nations and cities have sought economic success by encouraging different flavors of cosmopolitanism. Singapore became the wealthiest nation in Asia by inviting massive investment from far-flung and local trading partners. It is impossible to celebrate Singapore's success without acknowledging that Singapore is an authoritarian state where the same party has won every election for the past half century.[24] For better or worse, Singapore has emerged as a model for many states that seek economic growth by becoming a nexus of trade in goods and ideas.

Perhaps it's not a coincidence that the economic adviser who

helped lead the city-state's transformation was the Dutch economist Albert Winsemius. Winsemius visited Singapore on a UN
mission shortly after the nation achieved independence from
Britain, and became an unpaid adviser to the country for more
than twenty-five years. His advice covered matters large and
small, from inviting the Dutch electronics giant Phillips to build
manufacturing plants in the country to urging the government
not to remove a prominent statue of Stamford Raffles, the colonial leader who founded the city. Winsemius argued that the
statue of Raffles symbolized Singapore's openness to connection
with people from all nations, irrespective of previous colonial
relationships.[25]

What advice might a contemporary Winsemius offer a government intent on achieving economic success and development in
a connected world? Some choices, of course, would be tailored
to the level of development or government structure of a nation;
other principles might apply more generally. A Winsemius-
inspired briefing book would encourage connected (or would-be
connected) states to consider the following:

Physical Connection: Two key factors in Singapore's success
were the construction early on of a major container port and
an international airport with affordable landing rights. Nations
that have built an infrastructure that allows connection through
travel and trade also tend to be connected to the flow of people
and ideas. For the most disconnected nations, especially developing nations, strategies for connection begin with the basics of
infrastructure: roads and railroads, ports and airports, telephone
and power lines and Internet cables. For wealthier nations, questions of connection and disconnection loom as they decide how
to respond to security threats like terror and pandemic. The
temptation to close down physical connections to achieve mar-
ginal safety gains must be weighed against the disconnections
caused. And nations need to be cognizant of the gap between
their perception of their own international connectivity and how

connected they really are. Nations require an accurate picture of whom they are connected to and disconnected from, and why, as a first step toward becoming more central in the flow of goods and ideas.

Immigration: Highly connected nations are starting to see themselves as "market states,"[26] competing with other destinations for highly skilled immigrants. In response, they build immigration policies geared toward a highly mobile global workforce, creating more provisions for guest workers and workable paths toward citizenship for those who want it. They don't always get it right—the United Arab Emirates, for instance, will have to reform its bankruptcy laws to continue attracting European migrants. Its human rights and labor laws will also need an overhaul if the country is to continue attracting workers from the Middle East and India. And one of the secrets behind Singapore's economic success is a system of government that's far from open. Building a state that's both open and attractive as a market state is an unsolved problem.

Nations that aspire to close this gap between creating attractive markets and creating open societies need to pay special attention to immigrants who are positioned to build bridges between their former and their current homes. This might mean creating paths to citizenship for immigrants who come to a country to study and learn the local culture, since they're more likely to help fellow countrymen integrate and connect when they immigrate. And nations like the United States would do well to encourage immigrants with strong technical skills, who are likely to start new companies and transform old ones.

Education: Connected states need to invest heavily in education, preparing students to encounter a wide and complex world, both abroad and at home. Efforts to build world-class universities in Persian Gulf states reflect not just aspirations to be recognized as centers of learning as well as centers of wealth but also the wish to prepare citizens for their roles in internationally focused businesses, operating domestically or abroad. These universities are hiring from abroad not merely because more academic talent is available there, but because it creates spaces for intercultural encounter, like inviting the Jesuit priest Ryan Maher to teach theology to Qatari Muslim students at Georgetown's School of Foreign Service in Doha.[27]

In countries with well-established educational institutions, there's an opportunity to revamp education by putting the connected world at the center of a curriculum. Preparation for a connected world involves not just math and science but also language and geography education. Primary and secondary schools might focus on ways that students could connect with different groups of people in their local communities, embracing students and community members who are best positioned to bridge and translate between cultures and to encourage xenophilia as a core skill. Schools would be smart to mainstream study abroad programs.

Foreign Policy: In a connected world, trade and diplomacy are more powerful than they used to be. Infrastructures of connection can allow small groups to be disproportionately powerful, through direct action like terrorism or through the spread of disruptive and destructive information. At the same time, the transparency that connection brings makes it increasingly difficult to wage war without attracting international scrutiny, as America discovered at Abu Ghraib and as Israel has learned in the Gaza Strip. Winning wars is often easier than securing the peace, and military forces are transforming in some of the ways the military strategist Tom Barnett has predicted.[28] Such an army would feature a smaller force capable of defeating enemies and a much larger force dedicated to rebuilding and strengthening states, and it would employ as many anthropologists and linguists as explosives specialists.

A connected world encourages countries like the United States to massively expand public diplomacy projects like the Peace Corps. These programs would attempt to cultivate more bridge figures and xenophiles, and to identify and support the people who will work to repair the damage of war and build the peace in the future.

. . . .

IT'S HARD TO IMAGINE NATIONS in Europe or North America undertaking the changes mentioned above at a moment when economic slowdown has left nations looking to create jobs within their own borders. But connection is hard to undo, as the European Union has found out when its weak economies threaten the stability of its strong ones. In the long term, nations that want to benefit from connection need to start the long process of rewiring.

Some corporations, particularly large multinationals and some innovative technology companies, are engaged in a similar process. Like the transforming of a state, this transformation is neither quick nor easy. In certain cases it may not be feasible or desirable. But for companies trying to benefit from the cognitive diversity of their workforce, here's an ambitious set of steps to consider.

Recruit and Celebrate Bridge Figures: Smart multinationals create pathways for managers to work in different countries, spending years—not weeks—in different divisions and different countries. They're consciously cultivating bridge figures within their own organizations, increasing the chance that good ideas will be carried by the people closing the organization's structural holes. Organizations should consider the role of bridge figures when hiring. An employee who has lived and worked in different countries is likely to bring novel influences and perspectives into an organization, especially if it embraces the idea of learning from difference.

Cultivate Xenophiles: Not everyone is, or can be, a bridge between cultures. Corporations need people to cross these bridges as well. Outside the corporation, xenophiles often get interested in other countries through a cultural product that crosses borders: sports, music, film, or food. Within a corporation, xenophiles may build connections around shared projects or professional goals, but it's at least as likely that personal and cultural connection will build these ties. In a connected company, managers should find ways for members of their team to work on culturally diverse teams and, preferably, to travel to other countries and markets where the company builds or sells products. And corporations would be wise to invest in spaces—physical and virtual—where employees can connect, even if those connections are as far from ordinary work life as watching World Cup matches together or raiding in *World of Warcraft*.

Bridge and Translate Internally: Working in a global corporation doesn't guarantee anyone a global perspective, of course. Corporations need to look closely at the systems they use to share news within the corporation and bring in ideas and inspirations from outside. Monitoring a media diet isn't helpful just for individuals; it's important for managers and strategists to consider how they're learning about their company and their market. And it is as essential to build structures to allow translation of information and discovery of content within a company as it is to build better tools to transform global media.

Commit to the Long Haul: Bringing culturally diverse teams together doesn't ensure increased creativity or productivity—at first, there's likely to be more conflict than creativity. Companies brave enough to shift into a connected way of working will accept the short-term loss of productivity and work toward the long-term gains that come through cognitive diversity.

· · · · · · ·

NOT ALL OF US ARE in a position to rewire a country or a company. But all of us can take steps to increase the diversity of influences we're encountering and make stronger connections with perspectives.

Monitor Consumption: Self-tracking of the media we each consume is a first step toward understanding the biases we bring to the world. Maintaining a simple diary for a week is likely to be revelatory, while tools like RescueTime enable you to track your behavior over the long term, which is useful if monitoring turns into an effort to change your behavior.

Escape Your Orbit Slowly: If you discover that you spend a great deal of time consuming the same few types of media—as most of us do—you may be tempted to try to change your media diet all at once. A better first step is to pursue an interest you already have and look for international connections within that space. Whether it means following your interest in economics to read a Ugandan economist's blog or pursuing an interest in sumo to learn more about Mongolia, you're more likely to change your habits personally if you're following a topic that already fascinates you.

Find and Follow Bridge Figures: The best introduction to another country or culture is someone who understands that culture, and yours as well. The Internet is filled with people passionate about explaining their home cultures; our site Global Voices (globalvoicesonline.org) features many such individuals, but countless others exist. Communities like Meedan, sites like Tea Leaf Nation, and tools like Härnu all endeavor to introduce you to bridge figures who can help you understand another cultural context.

Seek Serendipity through Curators: Taking conscious steps toward diversifying the media you consume will take you only so far. We need to stumble on unexpected influences to make novel connections. This means granting some of our attention to curators—human and mechanical—who can introduce us to unexpected influences. Curators include editors of newspapers and literary magazines, new media curators like Maria Popova of Brain Pickings, and semiautomated systems like StumbleUpon and Longreads. In every case, seeking serendipity means embracing risk, being willing to let a curator lead you astray in exchange for moments of discovery.

These steps aren't as easy to take as they should be. In each instance, we would benefit from better tools to track and visualize what we're encountering, to make new connections within our fields of interest, and to find guides to new ideas and perspectives. There's an incredible opportunity to create tools that help people move beyond search and social modes of discovery and increase the chances of serendipity. Builders of these tools have a chance not just to gain great fiscal success but to make a positive impact on the world, increasing the range of perspectives and strategies we can bring to bear on complex problems.

The Task Ahead

The cultural critic Evgeny Morozov recently turned his fierce intellect toward the influential TED conference, arguing among other things that "since any meaningful discussion of politics is off limits at TED, the solutions advocated by TED's techno-humanitarians cannot go beyond the toolkit available to the scientist, the coder, and the engineer."[29]

This book focuses on a set of technological tools, the media, including news media, social media, and arts and culture. These tools shape our experience of the world, especially our experience of the world beyond our personal encounters with it. Like any technologies, media tools embody political assumptions made by their creators, consciously or otherwise. Facebook privileges connections to people you already know over people you might want to know. An online newspaper privileges your desire for the news you want over the unexpected news an editor shows you.

Recognizing the politics embedded in our media technologies and working to change them doesn't guarantee that corporations will value diverse perspectives of employees, or that nations will reconsider their immigration policies or diplomatic strategies. Choosing a world that favors connection over disconnection is a set of political battles, large and small, fought everywhere from the scale of national politics to the decisions we make as individuals when we open our web browsers.

The scientists whom Morozov excoriates for seeking a technological shortcut to problems of climate change are opportunists, in the best sense of the word. They see how new ways to harness solar energy or to produce biofuels from algae could make it easier for governments and individuals to undertake crucial changes that are vast in scale and politically hard to accomplish. Seizing an opportunity for technical leverage doesn't mean ignoring or escaping politics. Often, it's a way of changing the political balance of power around an issue.

Our broadcast news media are decades or centuries old, but they're experiencing a moment of rapid and disruptive change. This gives those who would like to rewire news to be more representative, more global, and more surprising, an opportunity. A similar disruption is taking place in the worlds of film and music, and rippling into other cultural spaces. The tools of social media that shape what we see and what we pay attention to are barely formed, and they change week to week, not year to year. If we are excited by the possibility of creating media that expose us to a wide range of perspectives, we have the opportunity to build the tools we need.

It's a mistake to assume that the Internet will inexorably bring about a connected future. But it's unhelpful to dismiss the ambitions of technological optimists like Howard Rheingold, Marconi, and Tesla simply because the futures they hoped for haven't yet come to pass. Their words might be read instead as prophecy. Rabbi Abraham Heschel, a leading religious scholar and civil rights leader, began his academic career with a vast study of biblical prophecy. While "prophecy," in modern parlance, has become associated with forecasting the future, in biblical times, prophets brought God's voice to the people to encourage them to change: "The prophet was an individual who said No to his society, condemning its habits and assumptions, its complacency, waywardness, and syncretism."[30] Read as prediction, the hopes of Marconi, Tesla, and Rheingold are clearly wrong. Read as prophecy, they challenge us to take control of our technologies and use them to build the world we want rather than the world we fear.

If we want a world that values diversity of perspective over the certainty of singular belief, a world where many voices balance a privileged few, where many points of view complicate issues and push us toward novel solutions, we need to build that world. To quote another rabbi, the first-century CE Rabbi Tarfon, "It's not incumbent on us to finish the work, but neither are we free to refrain from beginning it."[31] Whether we transform our own behavior, the tools we use to encounter the world, or our society as a whole, we have an opportunity to start the process of rewiring the world.

ACKNOWLEDGMENTS

Rewire is my first book. Like many first-time authors, I discovered that it is far easier to talk about writing a book than to actually complete one. It has been a long journey from idea and intention to the book you're reading, and without the help of dozens of friends, I would never have completed the voyage.

My agent, David Miller, was profoundly generous with his wisdom and expertise, helping me shape my ideas into something I could share with the rest of the world. Brendan Curry, my editor at W. W. Norton, is the answer to a first author's prayers. I am grateful for his help in smoothing the rough edges of my writing and in guiding me through the process of turning a manuscript into a book.

The Berkman Center for Internet and Society at Harvard University was my intellectual home while I wrote the majority of this book. I am grateful to John Palfrey for inviting me, almost a decade ago, to join this stimulating, challenging, and convivial community. I've learned a great deal from Berkman's directors: John Palfrey, Charles Nesson, Urs Gasser, Terry Fisher, Jonathan Zittrain, and, especially, Yochai Benkler, who has been an invaluable sounding board as I've worked through these ideas.

I'm grateful to the dozens of fellows I've had the chance to work with, especially Dan Gillmor, Andrew McLaughlin, Rebecca MacKinnon, Hal Roberts, Aaron Shaw, and Persephone Miel, of blessed memory. A special thanks is due to Dave Winer, who years ago convinced me that I should try blogging, my first step down a slippery slope that has led to more serious crimes, including authorship.

For me, the best part of Berkman has been the "book club," a group that has met weekly for years to help scholars turn their ideas into books. My fellow clubbers of books have become some of my closest colleagues and dearest friends, and I thank them all: danah boyd, Sasha Costanza-Chock, Judith Donath, Eszter Hargittai, Jason Kaufman, Colin Maclay, Christian Sandvig, Doc Searls, Wendy Seltzer, Lokman Tsui, and Zeynep Tufekçi. I owe special thanks to David Weinberger, who has honored the rest of us by treating us as peers while sharing the many lessons he has learned in writing his rich, intricate, and challenging books.

I completed this book at my new intellectual home, the MIT Media Lab. I am grateful to Mitch Resnick, Pattie Maes, and Nicholas Negroponte for inviting me to join their community, and thankful for Joi Ito's friendship and support as we both discover this legendarily fascinating institution. James Paradis has ensured that I feel as much at home in the Comparative Media Studies department at MIT as I do at the Media Lab, and I'm grateful for his close reading of a draft of this book. My students have contributed more than they know, both in shaping my thinking and in helping me learn how to communicate these ideas. I'm especially grateful to Nathan Matias, Matt Stempeck, and Molly Sauter for their comments on the manuscript, and offer special thanks to Molly, who did much of the hard work of footnote wrangling. Thanks to Rahul Bhargava for his work in creating two key illustrations, and to Lorrie LeJeune for her wisdom and friendship.

If Harvard and MIT have been my intellectual homes for the past decade, Global Voices has been my spiritual home and my extended family. I'm humbled by the work this extraordinary community does every day to share glimpses and impressions of the wider world with a wider audience. Thanks to everyone involved in the Global Voices family, and especially to Amira Al Hussaini, Boris Anthony, Sami ben Gharbia, Hossein Derakhshan, Onnik Krikorian, Solana Larsen, Gilad Lotan, Ndesanjo Macha, Georgia Popplewell, Lova Rakotomalala, David Sasaki, Ivan Sigal, Jillian York, and Portnoy Zheng, each of whom has challenged me to think through the issues of cosmopolitanism and connection. Global Voices would not exist without Rebecca MacKinnon, my cofounder, partner, and dear friend, and mere thanks seem inadequate to recognize the impact on my life, and the lives of many others, she has had by bringing this project to life.

I am grateful to the TED community for giving me a stage to share some of my ideas while I was early in the process of writing this book. Thanks to June Cohen, Chris Anderson, and Bruno Giussani for their friendship and support.

Numerous friends and colleagues have been kind enough to engage with me as I've worked through these ideas. An incomplete list of inspiring interlocutors includes Akwe Amosu, Kwame Anthony Appiah, Genevieve Bell, Joshua Benton, Ed Bice, John Bracken, Andy Carvin, Kate Crawford, Darius Cuplinskas, Cyrus Farivar, Shannon Farley, Henry Farrell, Pankaj Ghemawat, Brooke Gladstone, Josh Glenn, Margaret Gould Stewart, Allen Gunn, Matt Harding, Janet Haven, Erik Hersman, Ahmad Humeid, Alberto Ibargüen, Sherrilyn Ifill, Susan Landau, Zhang Lei, Marc Lynch, Cameron Marlow, Wayne Marshall, Alisa Miller, Jim Moore, Pippa Norris, Quinn Norton, Danny O'Brien, Dick O'Neill, Ory Okolloh, Eli Pariser, Xiao Qiang, István Rév, Howard Rheingold, Jay Rosen, Roland Soong, George Soros, Jonathan Stray, Cass Sunstein, Clive Thompson, Jenny Toomey, Johan Ugander, Katrin Verclas, and Nasser Weddady. Apologies to the valued friends I've failed to name here.

I did not understand the full weight of the word "favor" until I asked a set of friends to read a draft of this book and offer their thoughts. I'm deeply indebted to danah boyd, Judith Donath, Nathan Kurz, Rebecca MacKinnon, Nathan Matias, James Paradis, Molly Sauter, Clay Shirky, Matt Stempeck, Zeynep Tufekçi, and David Weinberger, and I promise to return the favor if any of them are kind enough to ask.

Thanks to Emily, Daniel, Chris, Amy, Seth, Debbie, Hank, Kate, Shannon, Jay, Sandy, Katherine, Colin, Nate, Emmy, Josh, Maggie, Embly, Mike, and the rest of the Noppet tribe for helping keep me sane, and keeping Rachel and Drew company.

My late mentor Dick Sabot taught me that the question of inequality—of resources, opportunity, or attention—was one of the most important topics one could study. I hope he would approve of how I've tried to address those questions, and I hope I've honored his memory with this book.

I come by my xenophilia honestly, because my parents, Don and Donna Zuckerman, have been fascinated by the width of the world since long before my birth. Thanks to them and my sister Liz for encouraging my wanderlust and always welcoming me home.

And finally, thanks to my wife, my best friend and my first reader, the Velveteen Rabbi, Rachel Barenblat. I am grateful for your patience, tolerance, and wisdom, and I love you more than I can say.

NOTES

INTRODUCTION

1. Amir Taheri, *The Spirit of Allah: Khomeini and the Islamic Revolution* (Bethesda, MD: Adler and Adler, 1986), pp. 17–18, 213.

2. Ibid., p. 18

3. The number of deaths at the Qom protests is widely disputed, with sources reporting between two and seventy dead. See Charles Kurzman, *The Unthinkable Revolution in Iran* (Cambridge: Harvard University Press, 2005), p. 37, and Ervand Abrahamian, *A History of Modern Iran* (Cambridge: Cambridge University Press, 2008), p. 158.

4. Abbas Milani, "Iran's Islamic Revolution: Three Paradoxes," openDemocracy, February 9, 2009, http://www.opendemocracy.net/article/iran-s-islamic-revolution-three-paradoxes.

5. Abrahamian, *History*, p. 161.

6. "Remembering Iran's 1979 Islamic Revolution," *Morning Edition*, NPR program hosted by Steve Inskeep, August 17, 2009, http://www.npr.org/templates/story/story.php?storyId=111944123.

7. Taheri, *Spirit*, chap. 1.

8. From an August 1978 CIA report quoted by Gary Sick, principal White House aide on Iran during the revolution, in his book *All Fall Down: America's Tragic Encounter with Iran* (New York: Random House, 1985), p. 92.

9. Bruce D. Berkowitz and Allan E. Goodman, *Best Truth: Intelligence in the Information Age* (New Haven: Yale University Press, 2002).

10. Susan Landau, *Surveillance or Security: The Risks Posed by New Wiretapping Technologies* (Cambridge: MIT Press, 2010), sec. 9.6.

11. Annabelle Sreberny-Mohammadi and Ali Mohammadi, *Small Media, Big Revolution: Communication, Culture, and the Iranian Revolution* (Minneapolis: University of Minnesota Press, 1994).

12. Landau, *Surveillance*, p. 216.

Chapter 1: CONNECTION, INFECTION, INSPIRATION

1. Chris Taylor, "The Chinese Plague," *World Press Review* 50 (2003), http://www.worldpress.org/Asia/1148.cfm.

2. Tim Spanton, "World's Deadliest Cough," *Sun*, July 28, 2007, http://www.thesun.co.uk/sol/homepage/news/158143/Worlds-deadliest-cough.html.

3. "SARS FAQ," Disaster Center, http://www.disastercenter.com/Severe%20Acute%20Respiratory%20Syndrome.htm.

4. Simon More, "Severe Acute Respiratory Syndrome (SARS)" (PowerPoint presentation at EADGENE Workshop on Interpreting Field Disease Data, Edinburgh, Scotland, February 1–2, 2010).

5. Donald J. MacNeil Jr., "Disease's Pioneer Is Mourned as a Victim," *New York Times*, April 8, 2003, http://www.nytimes.com/2003/04/08/science/disease-s-pioneer-is-mourned-as-a-victim.html.

6. Laurie Garrett, "Outbreak," review of *China Syndrome: The True Story of the 21st Century's First Great Epidemic*, by Karl Taro Greenfeld, in *Washington Post*, April 9, 2006, http://www.washingtonpost.com/wp-dyn/content/article/2006/04/06/AR2006040601666.html.

7. Alexander Batalin, "SARS Pneumonia Virus, Synthetic Manmade, according to Russian Scientist," Global Research, November 10, 2003, http://globalresearch.ca/articles/BAT304A.html.

8. World Health Organization, *SARS: How a Global Epidemic Was Stopped* (Geneva: WHO Press, 2006), chap. 15.

9. "Summary Table of SARS Cases by Country, 1 November 2002–7 August 2003," World Health Organization, August 15, 2003, http://www.who.int/csr/sars/country/country2003_08_15.pdf.

10. Jeffery Taubenberger and David Morens, "1918 Influenza: The Mother of All Pandemics," *Emerging Infectious Diseases* 12 (2006), http://wwwnc.cdc.gov/eid/article/12/1/05-0979_article.htm.

11. World Health Organization, *SARS*, introd.

12. "SARS Whistle-blower Breathing a Sigh of Relief," *China News Daily*, May 21, 2003, http://english.peopledaily.com.cn/200305/21/eng20030521_117004.shtml.

13. Paul Dourish and Genevieve Bell, *Divining a Digital Future: Mess and Mythology in Ubiquitous Computing* (Cambridge: MIT Press, 2011), p. 33.

14. Rania Abouzeid, "Bouazizi: The Man Who Set Himself and Tunisia

on Fire," *Time*, January 21, 2011, http://www.time.com/time/magazine/article/0,9171,2044723,00.html.

15. David Kirkpatrick, "Tunisia Leader Flees and Prime Minister Claims Power," *New York Times*, January 14, 2011, http://www.nytimes.com/2011/01/15/world/africa/15tunis.html; Robert Mackey's blog, *The Lede* (blog), *New York Times*, references Bouazizi two days earlier: "Tunisians Document Protests Online," January 12, 2011, http://thelede.blogs.nytimes.com/2011/01/12/tunisians-document-protests-online/.

16. Sal Gentile, "Octavia Nasr: US Media Missed 'the Anatomy' of Tunisia's Revolution," *PBS Need to Know*, January 21, 2011, http://www.pbs.org/wnet/need-to-know/the-daily-need/octavia-nasr-u-s-media-missed-the-anatomy-of-tunisias-revolution/6668/.

17. Kimberly Dozier, "Intelligence Community under Fire for Egypt Surprise," Associated Press, http://www.msnbc.msn.com/id/41423648/ns/politics-more_politics/t/intelligence-community-under-fire-egypt-surprise/#.UGNE_Pl27_A.

18. Pippa Norris summarizes research on international news in the late 1960s and 1970s, finding estimates of international and foreign news at between 25 percent and 40 percent of total news broadcast, in "The Restless Searchlight: Network News Framing of the Post Cold-War World," http://www.hks.harvard.edu/fs/pnorris/Acrobat/Restless%20Searchlight.pdf. Alisa Miller, citing work from Project for Excellence in Journalism, sees 10 percent foreign and international coverage in recent television broadcasts, in *Media Makeover: Improving the News One Click at a Time* (TED Books: 2011), Kindle ed.

19. "Mobile Phone Access Reaches Three Quarters of Planet's Population," World Bank press release, July 17, 2012, http://www.worldbank.org/en/news/2012/07/17/mobile-phone-access-reaches-three-quarters-planets-population.

20. Duncan J. Watts, *Six Degrees: The Science of a Connected Age* (New York: W. W. Norton, 2004), p. 283.

21. The reasons for Diogenes's exile are matters of historical dispute. One account of his exile states that he was banished for "defacing the currency." Historians are still trying to figure out precisely what this means. It's unclear whether Diogenes and his father, who may have been the treasurer of Sinope, were stealing money, or whether Diogenes defaced currency as a philosophical act of defiance.

22. Diogenes Laërtius, *Lives and Opinions of Eminent Philosophers*, trans. C. D. Bonge (London: Henry Bohn, 1852).

23. "The World Goes to Town," *Economist*, May 3, 2007, http://www.economist.com/node/9070726.

24. Margaret C. Jacob, "The Cosmopolitan as a Lived Category," *Daedalus* 137, no. 3 (Summer 2008): 18–25.

25. Robert D. Putnam, *"E Pluribus Unum*: Diversity and Community in the Twenty-First Century: The 2006 Johan Skytte Prize Lecture," *Scandinavian Political Studies* 30 (2007): 137–74.

26. Ibid.

27. Kwame Anthony Appiah, *Cosmopolitanism: Ethics in a World of Strangers* (New York: W. W. Norton, 2007), p. xv.

28. While this makes Appiah sound like a moral relativist, he defends himself from the charge by arguing for universal values that are shared across cultures, if obscured by "taboos" that are local in scope and application.

29. "Picasso's African-Influenced Period: 1907 to 1909," PabloPicasso .org, http://www.pablopicasso.org/africanperiod.jsp.

30. Andrew Meldrum, "Stealing Beauty: How Much Did Picasso's Paintings Borrow from African Art?," *Guardian*, March 14, 2006, http://www .guardian.co.uk/artanddesign/2006/mar/15/art.

31. Corinna Lotz, in *Apollo: The International Magazine for Collectors*, December 2007, p. 122, http://www.apollo-tmagazine.com/reviews/ books/386241/art-that-scared-picasso.thtml.

32. Meldrum, "Stealing Beauty."

33. Ibid.

34. *Picasso and Africa*, ed. Laurence Madeline and Marilyn Martin (Johannesburg: Standard Bank Gallery, 2006), catalog of an exhibition held at the Standard Bank Gallery, Johannesburg, February 10 through March 19, 2006, and the Iziko South African National Gallery, Cape Town, April 13 through May 21, 2006.

35. Léopold Sédar Senghor, "Masque nègre," in *Chants d'ombre* (1945), http://philosophie-et-litterature.oboulo.com/chants-ombre-leopold-sedar -senghor-1945-masque-negre-114650.html, cited in Lotz, "Art That Scared Picasso."

36. Roger E. Bohn and James E. Short, "How Much Information? 2009 Report on American Consumers," Global Information Industry Center of the University of California–San Diego, January 2010, http://hmi .ucsd.edu/pdf/HMI_2009_ConsumerReport_Dec9_2009.pdf.

37. Seventy minutes a day, according to the Pew Research Center, "Americans Spending More Time Following the News," September 12, 2010, http://people-press.org/2010/09/12/americans-spending-more-time -following-the-news/.

38. Howard Rheingold, *The Virtual Community: Homesteading on the Electronic Frontier*, rev. ed. (Cambridge: MIT Press, 2000), p. 181.

39. Tom Standage, *The Victorian Internet: The Remarkable Story of the Telegraph and the Nineteenth Century's On-line Pioneers* (New York: Berkley Books, 1999), p. 83.

40. Joseph J. Corn, *The Winged Gospel: America's Romance with Aviation* (Baltimore: Johns Hopkins University Press, 2002), p. 37.

41. Ivan Nardodny, "Marconi's Plans for the World," *Technical World Magazine* 18 (1912): 145–50.

42. From a 1926 interview in *Colliers*, quoted in "Marshall McLuhan Foresees the Global Village," http://www.livinginternet.com/i/ii_mcluhan.htm.

43. Langdon Winner, "Sow's Ears from Silk Purses: The Strange Alchemy of Technological Visionaries," in *Technological Visions: The Hopes and Fears That Shape New Technologies*, ed. Marita Sturken et al. (Philadelphia: Temple University Press, 2004), p. 34.

44. Personal correspondence with Howard Rheingold.

45. Benjamin Disraeli, *Vivian Grey* (London: Henry Colburn, 1826), bk. 6, chap. 7.

46. Jens Eric Gould, "The Making of the Innocence of Muslims: One Actor's Story," *Time*, September 13, 2012, http://nation.time.com/2012/09/13/the-making-of-innocence-of-muslims-one-actors-story.

47. Jones has gone on to burn copies of the Quran, though his actions have attracted less attention than his 2010 threat. See Kevin Sieff, "Florida Pastor Terry Jones's Koran Burning Has Far-reaching Effect," *Washington Post*, April 2, 2011, http://www.washingtonpost.com/local/education/florida-pastor-terry-joness-koran-burning-has-far-reaching-effect/2011/04/02/AFpiFoQC_story.html.

48. Pamela Constable, "Egyptian Christian Activist in Virginia Promoted Video That Sparked Furor," *Washington Post*, September 13, 2012, http://www.washingtonpost.com/local/egyptian-christian-activist-in-virginia-promoted-video-that-sparked-furor/2012/09/13/f2a52a4c-fdc4-11e1-b153-218509a954e1_story.html.

49. Robert Mackey and Liam Stack, "Obscure Film Mocking Muslim Prophet Sparks Anti-U.S. Protests in Egypt and Libya," *The Lede* (blog), *New York Times*, September 11, 2012, http://thelede.blogs.nytimes.com/2012/09/11/obscure-film-mocking-muslim-prophet-sparks-anti-u-s-protests-in-egypt-and-libya/; John Hudson, "The Egyptian Outrage Peddler Who Sent Anti-Islam YouTube Clip Viral," *Atlantic Wire*, September 13, 2012, http://www.theatlanticwire.com/global/2012/09/egyptian-outrage-peddler-who-sent-anti-islam-youtube-clip-viral/56826/.

50. Yoni Bashan, "Arrests Made after Police Officers Injured at Anti-Islamic Film Protest in Sydney CBD," *Daily Telegraph*, September

16, 2012, http://www.dailytelegraph.com.au/news/police-use-pepper -spray-on-anti-islamic-film-protesters-in-sydney-at-the-us-consulate /story-e6freuy9-1226474744811; "Belgian Police Detain 230 Protesting Anti-Islam Film," *Hürriyet Daily News*, September 16, 2012, http:// www.hurriyetdailynews.com/belgian-police-detain-230-protesting-anti -islam-film-.aspx?pageID=238&nID=30247&NewsCatID=351.

51. Mona Shadia and Harriet Ryan, "California Muslims Hold Vigil for Slain Ambassador," *Los Angeles Times*, September 15, 2012, http://articles .latimes.com/2012/sep/15/local/la-me-anti-muslim-film-20120915.

52. Judith S. Donath, "Identity and Deception in the Virtual Community," in *Communities in Cyberspace*, ed. Marc A. Smith and Peter Kollock (London: Routledge, 1999), pp. 27–58.

53. Marc Lynch, "The Failure of #MuslimRage," *Foreign Policy*, September 21, 2012, http://lynch.foreignpolicy.com/posts/2012/09/21/a_funny_ thing_happened_on_the_way_to_muslimrage.

54. Chris Stephen, "Bodies of Six Militiamen Found in Benghazi after Attacks on Bases," *Guardian*, September 22, 2012, http://www.guardian .co.uk/world/2012/sep/22/bodies-six-militiamen-found-benghazi ?newsfeed=true.

55. "People of Benghazi Trying to Save Chris Stevens Life before His Death," YouTube.com, September 17, 2012, http://www.youtube.com/ watch?v=hC0B0qrv2wA, http://www.youtube.com/watch?v=yMSnyOMRX os&feature=watch_response_rev.

56. "Mapping the Global Muslim Population," Pew Forum on Religion and Public Life, October 7, 2009, http://www.pewforum.org/Mapping -the-Global-Muslim-Population.aspx.

Chapter 2: IMAGINARY COSMOPOLITANISM

1. Nicholas Negroponte, *Being Digital* (New York: Knopf, 1995), pp. 3–4.

2. For more information on Fiji water, see the following: Rob Walker, "Water Proof," *New York Times*, June 1, 2008, http://www.nytimes.com /2008/06/01/magazine/01wwln-consumed-t.html; Claudia H. Deutsch, "For Fiji Water, a Big List of Green Goals," *New York Times*, November 7, 2007, http://www.nytimes.com/2007/11/07 /business/07fiji.html; Kerri Ritchie, "Fiji Media Told to Adopt 'Journalism of Hope,'" ABC (Australia), April 17, 2009, http://www.abc.net.au/news/2009-04-17/fiji-media -told-to-adopt-journalism-of-hope/1654410; Anna Lenzer, "Fiji Water: Spin the Bottle," *Mother Jones*, September/October 2009, http://www.mother jones.com/politics/2009/09/fiji-spin-bottle?page=2; Ed Dinger, "Fiji Water LLC," *Answers.com*, http://www.answers.com/topic/fiji-water-llc; AAP,

"Hollywood Couple Buys Fiji Water for $63m," *Sydney Morning Herald*, November 30, 2004, http://www.smh.com.au/news/Business/Hollywood -couple-buys-Fiji-Water-for-63m/2004/11/29/1101577419156.html.

3. John Maynard Keynes, *The Economic Consequences of Peace* (New York: Harcourt Brace & Howe, 1920), in paperback from Management Laboratory Press (2009), p. 20.

4. "U.S. at War: Globaloney," *Time*, February 22, 1943, http://www.time .com/time/magazine/article/0,9171,774367,00.html.

5. Jennifersharpe, "Denim Jeans," Sourcemap.com, http://sourcemap .com/view/478.

6. Andy Jerardo, "What Share of U.S. Consumed Food Is Imported?," *Amber Waves*, February 2008, http://webarchives.cdlib.org/sw1vh5dg3r/ http://ers.usda.gov/AmberWaves/February08/DataFeature/.

7. Bruce Blythe, "Wal-Mart's U.S. Grocery Sales Rose to Nearly $141 Billion Last Year," *Drover's Cattle Network*, April 1, 2011, http://www .cattlenetwork.com/cattle-news/company/Wal-Marts-US-grocery-sales -rose-to-nearly-141-billion-last-year-119064289.html.

8. "Steel Industry Executive Summary: July 2012," U.S. Department of Commerce, International Trade Administration, http://hq-web03.ita .doc.gov/License/Surge.nsf/webfiles/SteelMillDevelopments/$file/ exec%20summ.pdf?openelement; Mark Piepkorn, "Lumber by the Numbers," Continuing Education Center, http://continuingeducation .construction.com /article.php?L=5&C=645&P=1, originally published in *GreenSource*, January/February 2010 issue.

9. Daniel Cohen, *Globalization and Its Enemies*, trans. Jessica B. Baker (Cambridge: MIT Press, 2006), p. 52.

10. Jeffrey Frankel, "Globalization of the Economy," in *Governance in a Globalizing World*, ed. Joseph S. Nye and John D. Donahue (Washington, DC: Brookings Institution Press, 2000).

11. US Census, "Top Ten Countries with Which the United States Trades," May 2012, http://www.census.gov/foreign-trade/top/dst/current /balance.html.

12. Galina Hale and Bart Hobijn, "The U.S. Content of 'Made in China,'" Federal Reserve Bank of San Francisco Economic Letter, August 8, 2011, http://www.frbsf.org/publications/economics/letter/2011/el2011 -25.html?utm_source=home.

13. "Brazil's Victory in Cotton Trade Case Exposes America's Wasteful Subsidies," *Washington Post*, June 3, 2010, http://www.washington post.com/wp-dyn/content/article/2010/06/02/AR2010060204228.html; Will Allen, Eddie DeAnda, and Kate Deusterberg, "U.S. Cotton Subsidies: Killing Farmers & Poisoning Consumers & the Earth," *Organic Con-*

sumers Association, December 9, 2003, http://www.organicconsumers .org/clothes/willallen011504.cfm; Pietra Rivoli, *The Travels of a T-Shirt in the Global Economy* (Hoboken, NJ: John Wiley, 2009). Global cotton prices spiked well above subsidized levels in late 2010 and early 2011, which means subsidies weren't paid. But there have been few changes to the underlying subsidy system, and the United States will probably return to subsidizing cotton beyond this market spike.

14. Cohen, *Globalization*, p. 27.

15. Farzana Hakim and Colleen Harris, "Muslims in the European 'Media scape,'" Institute for Strategic Dialogue, 2009, http://www.strategicdialogue .org /ISD%20muslims%20media%20WEB.pdf.

16. "Facts and Figures," International Organization for Migration, http:// www.iom.int/jahia/Jahia/about-migration/facts-and-figures/lang/en.

17. These figures were calculated from data in the Wolfram Alpha database.

18. Adrian Michaels, "Muslim Europe: The Demographic Time Bomb Transforming Our Continent," *Telegraph*, August 8, 2009, http://www.telegraph .co.uk/news /worldnews/europe/5994047/Muslim-Europe-the-demographic -time-bomb-transforming-our-continent.html.

19. Karoly Lorant, "The Demographic Challenge in Europe," Brussels, April 2005, http://www.europarl.europa.eu/inddem/docs/papers/The%20 demographic%20challenge%20in%20Europe.pdf.

20. "The Future of the Global Muslim Population," Pew Forum on Religion and Public Life, January 27, 2011, http://www.pewforum.org/ The-Future-of-the-Global-Muslim-Population.aspx.

21. Federal Communications Commission, *International Telecommunications Data Report (2004)*, http://transition.fcc.gov/ib/sand/mniab/traffic/ files/CREPOR04.pdf, 1.

22. Data from International Telecommunication Union, http://www.itu .int/ITU-D/ict/statistics/.

23. Most Viewed Now, wwiTV.com, http://wwitv.com/most_viewed_now .htm.

24. All three publications offer large Chinese-language sections. Of course, these sites aren't always accessible to all Chinese, because of China's aggressive and ever-changing Internet censorship.

25. Nigeria ranked tenth in terms of total Internet users, but Doubleclick didn't track Nigeria at that point, so our set includes South Korea, then number eleven.

26. Asia and Japan Watch, the Asahi Shimbun, http://ajw.asahi.com/.

27. Priya Kumar, "Foreign Correspondents: Who Covers What?," *American Journalism Review*, December/January 2011, http://www.ajr.org/ article.asp?id=4997.

28. "Content Analysis," *The State of the News Media*, http://stateofthe media.org/2005/newspapers-intro/content-analysis/.
29. We plan to publish our study in early 2013. As in the Media Standards Trust study, we hand-coded every story in four American newspapers— the *New York Times*, the *Los Angeles Times*, the *Chicago Tribune*, and the *Washington Post*—for selected weeks separated by intervals of a decade. We saw major drops in international and foreign news stories in the *Chicago Tribune* and *Washington Post* over the forty-year period. The *Los Angeles Times* dropped sharply in both categories between 1979 and 1999, but had extensive international coverage in 2009; we are currently trying to understand if this is a data artifact, a week of heavy coverage for the paper, or a more general finding. We saw a major dropoff in foreign and international coverage in UK newspapers, but not in the *New York Times*, which was running 80 percent as many stories in these categories in 2009 as it was in 1979.
30. Guy Golan, "Inter-Media Agenda Setting and Global News Coverage: Assessing the Influence of the *New York Times* on Three Network Television Evening News Programs," *Journalism Studies* 7, no. 2 (2006): 323–33, http://syr.academia.edu/GuyJGolan/Papers/227056 /Inter-Media_Agenda_Setting_and_Global_News_Coverage_Assessing _the_Influence_of_The.
31. Alisa Miller, *Media Makeover: Improving the News One Click at a Time* (New York: Ted Books, 2011).
32. Kristen Purcell, Lee Rainie, Amy Mitchell, Tom Rosenstiel, and Kenny Olmstead, "Understanding the Participatory News Consumer," Pew Internet and American Life Project, March 1, 2010, http://www .pewinternet.org/Reports/2010/Online-News/Part-1/4-Satisfaction -with-coverage-of-different-news-topics.aspx.
33. "In a Changing News Landscape, Even Television Is Vulnerable," Pew Research Center, September 27, 2012, http://www.people-press .org/2012/09/27/section-3-news-attitudes-and-habits-2/.
34. Anthony Kaufman, "Is Foreign Film the New Endangered Species?," *New York Times*, January 22, 2006, http://www.nytimes.com/2006/01/22/ movies/22kauf.html.
35. Three Percent, http://www.rochester.edu/college/translation/three percent/index.php?s=about.
36. Amy Balkin, "In Transit," CabSpotting, http://cabspotting.org/ projects/intransit/; "FAQ," CabSpotting, http://cabspotting.org/faq.html.
37. "History of Railroads and Maps," Library of Congress, 1998, http:// memory.loc.gov/ammem/gmdhtml/rrhtml/rrintro.html; Olivier Zunz, *Making America Corporate* (London: University of Chicago Press, 1990).

38. Martin Dodge, "Cybergeography Research," University of Manchester, 2007, http://personalpages.manchester.ac.uk/staff/m.dodge/cyber geography/atlas/atlas.html.

39. Josh McWilliam, "How Does the Google Maps Traffic Feature Work? Where Does the Data Come From?," Bright Hub, September 24, 2010, http://www.brighthub.com/internet/google/articles/46896.aspx; Frederic Lardinois, "Google Maps Gets Smarter: Crowdsources Live Traffic Data," ReadWriteWeb, August 25, 2009, http://www.readwriteweb.com/archives /google_maps_gets_smarter_crowdsources_traffic_data.php.

40. Dave Barth, "The Bright Side of Sitting in Traffic: Crowdsourcing Road Congestion Data," *Googleblog*, August 25, 2009, http://googleblog .blogspot.com/2009/08 /bright-side-of-sitting-in-traffic.html.

41. Increasingly, operators also give this information to law enforcement officials when requested; the US mobile phone company Sprint provided 8 million records, many of which contained positioning information, in response to law enforcement requests between September 2008 and October 2009. See Kim Zetter's December 2009 article, "Feds 'Pinged' Sprint GPS Data 8 Million Times over a Year," *Wired*, http://www.wired.com/ threatlevel/2009/12/gps-data/.

42. Kai Biermann, "Tell-all Telephone," *Zeit Online*, March 26, 2011, http://www.zeit.de/datenschutz/malte-spitz-data-retention.

43. Noam Cohen, "Cellphones Track Your Every Move, and You May Not Even Know," *New York Times*, March 26, 2011, http://www.nytimes .com/2011/03/26/business/media/26privacy.html?_r=1.

44. Wiredautopia, "airtraffic," December 7, 2008, http://www.youtube .com/watch?feature=player_embedded&v=oR00_uLfGVE.

45. Research and Innovative Technology Administration, "Table 1-46: Air Passenger Travel Departures from the United States to Selected Foreign Countries by Flag of Carriers," in *National Transportation Statistics*, U.S. Dept. of Transportation, July 2010, http://www.bts.gov/publications /national_transportation_statistics /html/table_01_46.html.

46. Ibid.

47. Miller McPherson, Lynn Smith-Lovin, and James M. Cook, "Birds of a Feather: Homophily in Social Networks," *Annual Review of Sociology*, no. 27 (2001): 415–44, http://www.bsos.umd.edu/socy/alan/stats/ network-grad/handouts/McPherson-Birds%20of%20a%20Feather -Homophily%20in%20Social%20Networks.pdf.

48. Christian Jarrett, "We Sit Near People Who Look like Us," *BPS Research Digest* (blog), July 18, 2011, http://bps-research-digest.blogspot .com/2011/07/we-sit-near-people-who-look-like-us.html.

49. One group that showed the strongest homophily was students of

America's most elite boarding schools, providing ample fodder for conspiracies about an American ruling class.

50. George Simmel, *The Sociology of Georg Simmel*, trans. and ed. Kurt H. Wolff (Glencoe, IL: Free Press, 1950).

51. Andreas Wimmer and Kevin Lewis, "Beyond and Below Racial Homophily: ERG Models of a Friendship Network Documented on Facebook," *American Journal of Sociology* 116, no. 2 (September 2010): 583–642, http://www.sscnet.ucla.edu/soc/faculty/wimmer /WimmerLewis.pdf.

Chapter 3: WHEN WHAT WE KNOW IS WHOM WE KNOW

1. Robert F. Worth, "In New York Tickets, Ghana Sees Orderly City," *New York Times,* July 22, 2002, http://www.nytimes.com/2002/07/22/ nyregion/in-new-york-tickets-ghana-sees-orderly-city.html. In an ironic twist, the *New York Times* article that called attention to Ghana's role in NYC parking tickets led to scrutiny of the outsourcing decision, and the contract with the Ghanaian workers was not renewed.

2. "Op-Ed: An African Success Story," *New York Times*, January 8, 2001, http://www.nytimes.com/2001/01/08/opinion/an-african-success-story .html.

3. Peter Boyer, "Famine in Ethiopia," *Washington Journalism Review*, January 1985, pp. 18–21.

4. William C. Adams, "Whose Lives Count? TV Coverage of Natural Disasters," *Journal of Communication* 36 (Spring 1986): 113–22, http:// www.gwu.edu/~pad/202/readings/disasters.html.

5. This economic bias may become more pronounced in US news as Bloomberg and the *Wall Street Journal* are expanding their foreign coverage teams, while most other news organizations shrink their overseas footprint.

6. Maxwell E. McCombs and Donald L. Shaw, "The Agenda-Setting Function of Mass Media," *Public Opinion Quarterly* 36, no. 2 (Summer 1972): 176–87.

7. Ted Rall, "How the U.S. Media Marginalizes Dissent," *Al Jazeera*, August 4, 2011, http://english.aljazeera.net/indepth/opinion/2011/08 /20118164314283633.html.

8. Jay Rosen, "Audience Atomization Overcome: Why the Internet Weakens the Authority of the Press," *PressThink* (blog), January 12, 2009, http://archive.pressthink.org/2009/01/12/atomization.html.

9. Chris Roberts, "Gatekeeping Theory: An Evolution" (paper presented at Association for Education in Journalism and Mass Communica-

tion, San Antonio, TX, August 2005), http://www.chrisrob.com/about/gatekeeping.pdf.

10. Pamela Shoemaker and Tim P. Vos, *Gatekeeping Theory* (New York: Routledge, 2009), p. 16.

11. Dan Berkowitz, ed., *Social Meanings of News* (London: SAGE, 1997), p. 9.

12. MacBride commission, *Many Voices, One World: Toward a New, More Just, and More Efficient World Information and Communication Order* (Paris: UNESCO, 1980), p. 270.

13. Ibid., p. 263, rec. 47.

14. Ken Doctor, "The Newsonomics of WaPo's reader dashboard 1.0," *Nieman Journalism Lab*, April 7, 2011, http://www.niemanlab .org/2011/04/the-newsonomics-of-wapos-reader-dashboard-1-0/.

15. Michael Shapiro, "Six Degrees of Aggregation: How the Huffington Post Ate the Internet," *Columbia Journalism Review*, May/June 2012, http://www.cjr.org/cover_story/six_degrees_of_aggregation.php?page =all; "Google Analytics and Google Apps Help The Huffington Post Keep Its Edge," http://www.google.com/analytics/customers/case_study_ huffington_post.html.

16. William Safire, *Before the Fall: An Inside View of the Pre-Watergate White House* (New Brunswick, NJ: Transaction Publishers, 2005), p. 392.

17. Diana Saluri Russo, "Is the Foreign News Bureau Part of the Past?," *Global Journalist*, January 30, 2010, http://www.globaljournalist.org/ stories /2010/01/30 /is-the-foreign-news-bureau-part-of-the-past/.

18. Jodi Enda, "Retreating from the World," *American Journalism Review*, December/January 2011, http://www.ajr.org/article.asp ?id=4985.

19. *"The Wall Street Journal* under Rupert Murdoch," *Journalism.org*, July 20, 2011, http://www.journalism.org/commentary_backgrounder/ wall_street_journal_under_rupert_murdoch.

20. I asked the head of analytics at the *New York Times* whether he could share information on traffic to international stories on the paper versus domestic stories. He was polite, but firm in his refusal, explaining that the *Times* doesn't even share that information with advertisers.

21. Elizabeth A. Thomson, "Freshman Publishing Experiment Offers Made-to-Order Newspapers," *MITnews*, March 9, 1994, http://web.mit .edu/newsoffice/1994 /newspaper-0309.html.

22. Cass R. Sunstein, *Infotopia: How Many Minds Produce Knowledge* (New York: Oxford University Press, 2006), p. 188.

23. Eric Lawrence, John Side, and Henry Farrell, "Self-Segregation or Deliberation? Blog Readership, Participation, and Polarization in

American Politics," Department of Political Science, George Washington University, March 10, 2009, http://projects.iq.harvard.edu/cces/files/Lawrence_Sides_Farrell-_Self-Segregation_or_Deliberation.pdf.

24. Natalie Glance and Lada Adamic, "The Political Blogosphere and the 2004 U.S. Election: Divided They Blog," *Proceedings of the 3rd International Workshop on Link Discovery* (2005): 36–43.

25. Eszter Hargittai et al., "Cross-ideological Discussions among Conservative and Liberal Bloggers," *Public Choice* 134 (2008): 67–86.

26. John Horrigan, Kelly Garrett, and Paul Resnick, "The Internet and Democratic Debate," Pew Internet and American Life Project, October 27, 2004, http://www.pewinternet.org/Reports/2004/The-Internet-and-Democratic-Debate.aspx.

27. Mark Glaser, "Social Media Grows at NY Times, But Home Page Remains King," *MediaShift*, January 13, 2011, http://www.pbs.org/mediashift/2011/01 /social-media-grows-at-ny-times-but-home-page-remains-king 013.html.

28. Brian Stelter, "Finding Political News Online, the Young Pass It On," *New York Times*, March 27, 2008, http://www.nytimes.com/2008/03/27/us/politics/27voters .html?_r=1.

29. Dee T. Allsop, Bryce R. Bassett, and James A. Hoskins, "Word-of-Mouth Research: Principles and Applications," *Journal of Advertising Research*," December 2007, pp. 398–410.

30. Stelter, "Finding Political News Online."

31. Kenny Olmstead, Amy Mitchell, and Tom Rosenstiel, "Where People Go, How They Get There, and What Lures Them Away," Pew Research Center's Project for Excellence in Journalism, May 9, 2011, http://www.journalism.org/node /25008.

32. Alexis Madrigal, "Dark Social: We Have the Whole History of the Web Wrong," *Atlantic*, October 12, 2012, http://www.theatlantic.com/technology/archive/2012/10/dark-social-we-have-the-whole-history-of-the-web-wrong/263523/.

33. Justin Osofsky, "Making News and Entertainment More Social in 2011," *Facebook Developer Blog*, December 28, 2010, http://developers.facebook.com/blog/post/443/.

34. Tanya Cordrey, "Tanya Cordrey's Speech at the Guardian Changing Media Summit," *Guardian*, March 21, 2012, http://www.guardian.co.uk/gnm-press-office/changing-media-summit-tanya-cordrey.

35. "What Americans Do Online: Social Media and Games Dominate Activity," *Nielsen Wire* (blog), August 2, 2010, http://blog.nielsen.com/nielsenwire/online_mobile /what-americans-do-online-social-media-and-games-dominate-activity/; "Social Networking Accounts for 1 of Every 5 Min-

utes Spent Online in Australia," comScore, Inc., February 18, 2011, http://www.comscore.com/layout/set/popup/Press_Events/Press_Releases/2011/2/Social_Networking_Accounts_for_1_of_Every_5_Minutes_Spent_Online_in_Australia.

36. Matt Carmichael and E. J. Schultz, "Millennial Grocery Shopping Habits and Marketing Trends," *AdAge Blogs*, June 29, 2011, http://adage.com/article/adagestat/millennial-grocery-shopping-habits-marketing-trends/228480/.

37. Jason Kincaid, "The Secret Sauce That Makes Facebook's News Feed Tick," *TechCrunch*, April 22, 2010, http://techcrunch.com/2010/04/22/facebook-edgerank/.

38. Eli Pariser, *The Filter Bubble: What the Internet Is Hiding from You* (New York: Penguin Press, 2011), p. 5.

39. Barry Schwartz, "Duck! Google's Cutts Responds to Search Filter Bubbles," *Search Engine Roundtable*, June 21, 2011, http://www.seround table.com/google-search-bubble-response-13591.html.

40. Bill Bishop, "The Big Sort Maps," *The Big Sort*, http://www.thebig sort.com /maps.php.

41. Keith N. Hampton, Lauren Sessions Goulet, Lee Rainie, and Kristen Purcell, "Social Networking Sites and Our Lives," Pew Research Center's Internet and American Life Project, June 16, 2011, http://www.pewinter net.org/~/media//Files /Reports/2011/PIP%20-%20Social%20networking%20sites%20and%20our%20lives.pdf.

42. Garry Blight, Sheila Pulham, and Paul Torpey, "Arab Spring: An Interactive Timeline of Middle East Protests," *Guardian*, January 5, 2012, http://www.guardian.co.uk/world/interactive/2011/mar/22/middle-east-protest-interactive-timeline.

43. Ethan Zuckerman, "What Bloggers Amplify from the BBC," *my heart's in accra*, January 28, 2005, http://www.ethanzuckerman.com/blog/2005/01/28/what-bloggers-amplify-from-the-bbc/; "*New York Times* Headlines in Blogs, Seen by Technorati," http://gapdev.law.harvard.edu/results/20050128/nytheadlines.html.

44. Jonah Berger and Katherine L. Milkman, "What Makes Online Content Viral?" *Journal of Marketing Research* 49, no. 2 (2012): 192–205, http://www.journals.marketingpower.com/doi/abs/10.1509/jmr.10.0353.

45. Paul Butler, "Visualizing Friendships," Facebook's Engineering Posts, December 13, 2010, https://www.facebook.com/notes/facebook-engineering/visualizing-friendships/469716398919. One striking thing about Butler's map is what's missing—much of Asia. Facebook is blocked in China and has had difficulty gaining traction in South Korea and Japan.

46. Johan Ugander, Brian Karrer, Lars Backstrom, and Cameron Mar-

low, "The Anatomy of the Facebook Social Graph," November 18, 2011, http://arxiv.org/pdf/1111.4503v1.pdf. Discussions with the authors suggest that this paper overrepresents Canada because of a miscounting of Facebook users who use Blackberry devices, and the actual percentage of international ties is closer to 13.3 percent.

47. "Stories," http://www.facebookstories.com/stories/1574/interactive -mapping-the-world-s-friendships#color=continent&story=1&country =US.

48. Personal conversation via email, October 10, 2012.

Chapter 4: GLOBAL VOICES

1. "Global Voices Manifesto," *Global Voices Online*, April 9, 2007, http:// globalvoicesonline.org/about/gv-manifesto/.

2. Mahmood Nasser Al-Yousif, "About," *Mahmood's Den*, http://mahmood.tv/about/.

3. Jennifer Preston, "When Unrest Stirs, Bloggers Are Already in Place," *New York Times*, March 13, 2011, http://www.nytimes.com/2011/03/14/business /media/14voices.html.

Chapter 5: FOUND IN TRANSLATION

1. "Jay Walker on the World's English Mania," TED talk, February 2009, http://www.ted.com/talks/jay_walker_on_the_world_s_english_mania .html.

2. Edward T. O'Neill, Brian F. Lavoie, and Rick Bennett, "Trends in the Evolution of the Public Web," *D-Lib Magazine* 9, no. 4, April 2003, http://www.dlib.org /dlib/april03/lavoie/04lavoie.html.

3. Ted Smalley Bowen, "English Could Snowball on the Net," *Technology Research News*, November 21, 2001, http://www.trnmag.com /Stories/2001/112101/English_could_snowball_on_Net_112101.html; Neil Gandal and Carl Shapiro, "The Effect of Native Language on Internet Usage," November 12, 2001, http://econ.tau.ac.il/papers/applied /language.pdf.

4. Email exchange, February 3, 2011.

5. "Internet World Users by Language," *Internet World Stats*, http://www .internetworldstats.com/stats7.htm.

6. Marissa McNaughton, "Over 50% of China's Internet Users Regularly Blog and Use Social Media," *Realtime Report*, January 10, 2011, http:// twtrcon.com /2011/01/10 /over-50-of-chinas-internet-users-regularly-blog -and-use-social-media.

7. Ahmad Humeid, *360° East*, http://www.360east.com/.

8. Daniel Sorid, "Writing the Web's Future in Many Languages," *New York Times*, December 30, 2008, http://www.nytimes.com/2008/12/31/technology /internet/31hindi.html?pagewanted=2.

9. After all, she speaks only English, Spanish, Danish, German, and French.

10. A number of philosophers and linguists have offered the idea of considering language as a tool, rather than as an innate cognitive feature. See Andy Clark, "Magic Words: How Language Augments Human Computation," http://www.nyu.edu/gsas/dept/philo/courses/concepts/magic words.html, or Daniel L. Everett, *Language: The Cultural Tool* (New York: Pantheon, 2012).

11. Bertran C. Bruce, "The Disappearance of Technology: Toward an Ecological Model of Literacy," https://www.ideals.illinois.edu/bitstream/handle/2142/13343/disappearance_of_tech.pdf.

12. Eytan Adar et al., "Information Arbitrage across Multi-lingual Wikipedia," *Proceedings of the Second ACM International Conference on Web Search and Data Mining* (New York, 2009).

13. Robert K. Plumb, "Russian Is Turned into English by a Fast Electronic Translator," *New York Times*, January 8, 1954.

14. "Assembly language" is the "native language" that computers speak. It features a very limited set of instructions that a microprocessor can execute. Programming in assembly language is painstaking and difficult, much more difficult than programming in modern programming languages, which bear at least a vague resemblance to human language. These "high-level languages" weren't implemented until three years later, when Sheridan helped IBM develop FORTRAN.

15. W. John Hutchins, "The Georgetown-IBM Experiment Demonstrated in January 1954," http://www.hutchinsweb.me.uk/AMTA-2004.pdf.

16. Barry Newman, "U.S. Government Sells News Service with World News Gathered by the CIA," *Wall Street Journal*, February 28, 2011, http://online.wsj.com/article/SB100014240527487046290045761363811 78584352.html.

17. Interview of Soong via email, February 5, 2011.

18. Ethan Zuckerman, "Who's Writing about Lu Banglie?," *my heart's in accra* (blog), October 10, 2005, http://www.ethanzuckerman.com/blog/2005/10/10 /whos-writing-about-lu-banglie/; Roland Soong, "The Taishi (China) Elections—Part 1 (Chronology)," *EastSouthWestNorth* (blog), September 19, 2005, http://www.zonaeuropa.com/20050919_1.htm.

19. Current editions are available at http://www.ecocn.org/portal.php.

20. Yochai Benkler, *The Wealth of Networks: How Social Production*

Transforms Markets and Freedom (New Haven: Yale University Press, 2006), p. 83.

21. The TED conference moved from Monterey to Long Beach, California, in 2009, and TED talks online now include talks from the main conference, TED Global, hosted in the UK, and from TEDx talks around the world.

22. Luis von Ahn et al., "reCAPTCHA: HumanBased Character Recognition via Web Security Measures," *Science*, September 12, 2008, pp. 1465–68.

23. Luis von Ahn, "Massive-Scale Online Collaboration," TED talk, April 2011, http://www.ted.com/talks/luis_von_ahn_massive_scale_online_collabo ration.html.

24. Christopher Mims, "Translating the Web While You Learn," *MIT Technology Review*, May 2, 2011, http://www.technologyreview.com/news/423894/translating-the-web-while-you-learn/.

25. Philipp Koehn, "Europarl: A Parallel Corpus for Statistical Machine Translation" (paper presented at Machine Translation Summit, 2005), http://mt-archive.info/MTS-2005-Koehn.pdf.

26. J. M. Ledgard, "Digital Africa," *Intelligent Life*, Spring 2011, http://moreintelligentlife.com/content/ideas/jm-ledgard/digital-africa?page=full.

27. Wade Davis, "Dreams from Endangered Cultures," filmed February 2003, TED video, 22:05, posted January 2007, http://www.ted.com/talks/wade_davis_on_endangered_cultures.html.

Chapter 6: TAKEN IN CONTEXT

1. Marc Eliot, *Paul Simon: A Life* (Hoboken, NJ: John Wiley, 2010), p. 186.

2. The identity of the specific cassette Berg gave Simon is the subject of much speculation by music historians. It's widely reported the cassette was titled "Gumboots: Accordion Jive Hits, volume 2," but no one has been able to find a record released with that name. Frustrated and fascinated by the elusiveness of the source material, Eric Kleptone, a British musician who has built a career on remixing other people's source material, released a mix online called "Paths to Graceland," featuring music that might have been on the legendary cassette—http://www.kleptones.com/blog/2012/06/28/hectic-city-15-paths-to-graceland.

3. Timothy White, "Lasers in the Jungle: The Conception and Maturity of a Musical Masterpiece," Warner Bros. Records, 1997, http://www.wbr.com/paulsimon/graceland/cmp/essay.html.

4. The Boyoyo Boys, *All Music*, http://www.allmusic.com/artist/the-boyoyo-boys-mn0000625695.

5. "The Boyoyo Boys and Township Jive Today," *Samaka Music* (blog), December 9, 2007, http://samakamusic.blogspot.com/2007/12/boyoyo-boys-and-township-jive-today.html.

6. Simon is not a perfect person or a perfect xenophile; it's certainly possible to see his work as appropriation. The last two tracks on *Graceland* feature the Latino band Los Lobos, whose members are credited as musicians, not as songwriters. Steven Berlin of Los Lobos claims that "The Myth of Fingerprints" is a song he and his band wrote, titled "By Light of the Moon," and has threatened to sue Simon, calling him "the world's biggest prick, basically." Then again, Berlin hasn't sued Simon.

7. "Notes on Odysseus' Name and Pseudonyms," http://faculty.gvsu.edu/websterm/m%27onomakluton.html.

8. Alexis Bloom and Tshewang Dendup, "Bhutan's Busiest Cable Guy," *FRONTLINE/World*, December 2000, http://www.pbs.org/frontlineworld/stories/bhutan/interview.html.

9. Somini Sengupta, "Bhutan Lets the World In (But Leaves Fashion TV Out)," *New York Times*, May 6, 2007, http://www.nytimes.com/2007/05/06/world/asia/06bhutan.html.

10. Cathy Scott-Clark and Adrian Levy, "Fast Forward into Trouble," *Guardian*, June 13, 2003, http://www.guardian.co.uk/theguardian/2003/jun/14/weekend7 .weekend2.

11. Pat Harris, "City of Nashville rejects English-only law," *Reuters* online, January 23, 2009, http://www.reuters.com/article/2009/01/23/us-usa-english-nashville-idUSTRE50M11420090123.

12. Pippa Norris, *Cosmopolitan Communications: Cultural Diversity in a Globalized World* (Cambridge: Cambridge University Press. 2009), Kindle ed., locations 549–52.

13. Ibid., chap. 10.

14. Malcolm Gladwell, *The Tipping Point: How Little Things Can Make a Big Difference* (Boston: Little, Brown, 2000), p. 54.

15. "Why Americans Use Social Media," Pew Internet and American Life Project, http://www.pewinternet.org/Reports/2011/Why-Americans-Use-Social-Media/Main-report.aspx.

16. I'm grateful to my friend and colleague Zeynep Tufekçi for this observation and her help in interpreting Granovetter.

17. Ronald S. Burt, "Structural Holes and Good Ideas," *American Journal of Sociology* 110, no. 2 (September 2004): 349–99, http://www.bebr.ufl.edu/sites /default/files/Structural%20Holes%20and%20Good%20Ideas.pdf.

18. "Akademie Olympia," Albert Einstein in the World Wide Web, http://www.einstein-website.de/z_biography/olympia-e.html.

19. Burt, "Structural Holes."

20. Ibid.

21. If there's an ominous tone to Putin's request for top-lap's address—and therefore, his identity—that's probably not a coincidence. Five months after his post "went viral," his blog was discontinued. Top-lap's last few blog entries reported on a visit from the police, who had confiscated his hard drive and USB key, and signaled his intent to flee in the hopes of avoiding arrest.

22. Jay Rosen, "National Explainer: A Job for Journalists on the Demand Side of News," *PressThink* (blog), August 13, 2008, http://archive.press-think.org/2008/08/13/national_explain.html.

23. Tiffany Ap, "The Wild, Wile Web: Ever-Elusive, chinaSMACK Founder Fauna," *The Beijinger* (blog), July 26, 2010, http://www.thebeijinger.com/blog/2010/07/26/The-Wild-Wile-Web-Ever-Elusive-chinaSMACK-founder-Fauna; Maile Cannon and Jingying Yang, "Bloggers Open an Internet Window on Shanghai," *New York Times*, February 24, 2010, http://www.nytimes.com/2010/02/25/technology/25iht-rshanblog.html?pagewanted=all.

24. Fauna, "Migrant Workers' Children Spend Childhood Scavenging Landfill," chinaSMACK, August 24, 2011, http://www.chinasmack.com/2011/pictures/migrant-workers-children-spend-childhood-scavenging-landfill.html.

25. Human Library, "About," 2012, http://humanlibrary.org/the-history.html; Andrew Weichel, "Surrey Library to Loan Out Volunteers as 'Living Books,'" *CTV News*, August 20, 2011, http://www.ctvbc.ctv.ca/servlet/an/local/CTVNews.

26. Achal R. Prabhala, *People Are Knowledge* (video), posted July 15, 2011, http://vimeo.com/26469276; Noam Cohen, "A Push to Redefine Knowledge at Wikipedia," *New York Times*, August 7, 2011, http://www.nytimes.com/2011/08/08/business/media/a-push-to-redefine-knowledge-at-wikipedia.html?_r=1.

27. Dhani Jones and Jonathan Grotenstein, *The Sportsman* (New York: Rodale Books, 2011), Kindle ed., locations 3885–86.

28. Ethan Zuckerman, "Wayne Marshall on Nu Whirled Music . . . and My Thoughts, Too . . . ," *my heart's in accra* (blog), December 15, 2010, http://www.ethanzuckerman.com/blog/2010/12/15/wayne-marshall-on-nu-whirled-music-and-my-thoughts-too/.

29. David Drake, "Diplo: The Stylus Interview," *Stylus*, October 4, 2004, http://www.stylusmagazine.com/articles/weekly_article/diplo-the-stylus-interview.htm.

30. The term "baile funk" actually refers to the dances where funk is played—the music is often referred to as "funk carioca" in Brazil. But "baile funk" is generally the term used outside Brazil to discuss the music, and I'm following that convention here.

31. Sasha Frere-Jones, "Brazilian Wax," *New Yorker*, August 1, 2005, http://www.newyorker.com/archive/2005/08/01/050801crmu_music ?currentPage=1.

32. Camilo Rocha, "Globalistas: An Introduction," Spannered, February 20, 2008, http://www.spannered.org/music/1375/.

33. Matt Harding, *Where the Hell Is Matt?: Dancing Badly around the World* (New York: Skyhorse Publishing, 2009), p. 77.

34. Ibid., p. 143.

35. Hugo Zemp, "The/An Ethnomusicologist and the Record Business," *Yearbook for Traditional Music* 28 (1996): 36–56. See also Stephen Feld, "A Sweet Lullaby for World Music," *Public Culture* 12, no. 1 (2000): 145–71.

36. Matt Harding, "Auki, Solomon Islands Loose Ends," *Wherethehell ismatt.com*, February 4, 2011, http://www.wherethehellismatt.com/jour nal/2011/02/auki-solomon-islands.html.

Chapter 7: SERENDIPITY AND THE CITY

1. "Lagos Makoko Slums Knocked Down in Nigeria," *BBC News Africa*, July 17, 2012, http://www.bbc.co.uk/news/world-africa-18870511.

2. World Bank, "Urban Development," 2012, http://data.worldbank.org/ topic /urban-development.

3. John Grimond, "A Survey of Cities: The World Goes to Town," *Econo-mist*, May 3, 2007, http://www.economist.com/node/9070726; "European Urbanization" (map), *Philip's Atlas of World History*, http://qed.prince ton.edu/index.php/User:Student/European_Urbanization_1800; Popula-tion Reference Bureau, "Human Population: Urbanization," 2012, http:// www.prb.org/Educators/TeachersGuides/HumanPopulation/Urbaniza tion.aspx; *World Urbanization Prospects: The 2007 Revision* (New York: United Nations, 2007), http://www.un.org/esa/population/publications/ wup2007/2007WUP_Highlights_web.pdf.

4. It didn't help that the mayor of London had ordered all cats and dogs killed for fear they were spreading the plague—instead, they were likely keeping the plague rats in check.

5. David Perdue, "Dickens' London," David Perdue's Charles Dickens Page, 2012, http://www.fidnet.com/~dap1955/dickens/dickens_london .html; "John Snow: Broad Street Pump Outbreak," UNC Gillings School

of Global Public Health, http://courses.sph.unc.edu/john_snow/pro
logue.htm.

6. Martin Daunton, "London's 'Great Stink' and Victorian Urban Planning," *BBC History*, August 22, 2012, http://www.bbc.co.uk/history/
trail/victorian_britain /social_conditions/victorian_urban_planning_01
.shtml.

7. Ethan Zuckerman, "Geek Tracking, African Hacking," *my heart's in
accra* (blog), April 27, 2007 (2:08 pm), http://www.ethanzuckerman.com/
blog /2007/04/27/geek-tracking-african-hacking/.

8. Guy Debord, "Theory of the Dérive," trans. Ken Knabb, in *Situationist International Anthology* (2006), posted on http://www.bopsecrets.org/
SI/2.derive.htm.

9. Zachary M. Seward, "Everything the Internet Knows about Me
(Because I Asked It To)," *Digits* (blog), *Wall Street Journal*, December
22, 2010, http://blogs.wsj.com/digits/2010/12/22/everything-the-internet
-knows-about-me-because-i-asked-it-to/.

10. Ibid.

11. Brad DeLong, "What Future Does Facebook Have?," *Brad DeLong*
(blog), January 10, 2011, http://delong.typepad.com/sdj/2011/01/what
-future-does-facebook-have.html.

12. Paul Carr, "NSFW: #Ebony and #Ivory—The Brave New World of
Online Self-Segregation," TechCrunch, May 2, 2010, http://techcrunch
.com/2010/05/02/a-limey-writes/.

13. "YESTERDAY: The Rabbit Proof Fence," State Barrier Fence of
Western Australia, http://pandora.nla.gov.au/pan/43156/20040709-0000/
agspsrv34.agric.wa.gov.au/programs/app/barrier/history.htm.

14. Letter to Sir Horace Mann of January 28, 1754, cited in T. G. Remer,
Serendipity and the Three Princes, from the Peregrinaggio of 557 (Nor-
man: University of Oklahoma Press, 1965).

15. Colin Eagan, "What the Design of Cities Teaches Us about the Design
of Sites," Fit & Finish: Insight from the ICF Ironworks User Experience
Group, July 27, 2010, http://fitandfinish.ironworks.com/2010/07/what
-designing-cities-teaches-us-about-designing-sites-or-how-not-to
-become-the-digital-equivalent-of-tysons-corner.html.

16. The Library of Congress announced plans to archive all public Twitter messages in 2010. By 2012, those messages were not accessible, and
reporting on the project focused on the technical challenges the Library
of Congress was experiencing in managing the large data set Twitter creates: http://radar.oreilly.com/2011/06/library-of-congress-twitter-archive
.html.

17. Matthew Ingram, "Which Is Better, Real Names on Facebook or Helping

Dissidents?," GigaOm, February 8, 2011, http://gigaom.com/2011/02/08/which-is-better-real-names-on-facebook-or-helping-dissidents/.

18. Jennifer Woodard Maderazo, "Facebook Becomes Catalyst for Causes, Colombian FARC Protest," *MediaShift*, February 22, 2008, http://www.pbs.org/mediashift/2008/02/facebook-becomes-catalyst-for-causes-colombian-farc-protest053.html.

19. The Pages Directory, ranked by page popularity, is accessible here: http://www.facebook.com/directory/pages/.

20. Gilad Lotan, "Data Reveals That 'Occupying' Twitter Trending Topics Is Harder Than It Looks!," *Social Flow* (blog), October 12, 2011, http://blog.socialflow.com/post/7120244374/data-reveals-that-occupying-twitter-trending-topics-is-harder-than-it-looks.

21. Jon Fasman, "The $10,000 Gin and Tonic," *Intelligent Life*, September 2007, http://moreintelligentlife.com/node/153.

22. Guy Debord, "Theory of the Dérive," trans. Ken Knabb, in *Situationist International Anthology* (2006), posted on http://www.bopsecrets.org/SI/2.derive.htm.

23. Linda Monach, *Burgers Here and There*, 2011, http://burgershereandthere.com/.

24. David Weinberger, "Free Dewey!," *KMWorld*, October 1, 2004, http://www.kmworld.com/Articles/Column/David-Weinberger/Free-Dewey!-9579.aspx.

25. S. Roberts, "Self-experimentation as a Source of New Ideas: Examples about Sleep, Mood, Health, and Weight," *Behavioral and Brain Sciences* 27 (2004): 227–62, http://quantifiedself.com/2011/03/effect-of-one-legged-standing-on-sleep/.

Chapter 8: THE CONNECTED SHALL INHERIT

1. The Sopranos, "Made in America," *The Sopranos*, June 10, 2007.

2. "Journey Announces New Singer," *Blabbermouth.net*, December 5, 2007, http://www.blabbermouth.net/news.aspx?mode=Article&newsitemID=86205.

3. "Arnel Pineda Biography," Biography.com, http://www.biography.com/people/arnel-pineda-20860785.

4. Benito "Sunny" Vergara outlines the idea of *plakado* and Arnel Pineda's improbable journey in a set of blog posts: http://americanpop.asianweek.com/2008/06/tongues-like-parrots/, http://americanpop.asianweek.com/2008/07/the-man-can-sing-anything/, http://americanpop.asianweek.com/2008/07/its-steve-and-its-not-steve/.

5. Alex Pappademas, "He Didn't Stop Believin'," *GQ*, June 2008, http://www .gq.com/entertainment/music/200805/arnel-pineda-journey-lead-singer.

6. Patricia Sellers, "It's Good to Be the Boss," *CNN Money*, October 2, 2006, http://money.cnn.com/2006/09/29/magazines/fortune/mpw.female CEOs.intro.fortune/index.htm.

7. Rob Cox, "Will Indra Nooyi Leave Pepsi?," *DailyBeast.com*, July 30, 2012, http://www.thedailybeast.com/newsweek/2012/07/29/will-indra-nooyi -leave-pepsi.html.

8. "Senior Functional Leadership: Muhtar Kent," Coca-Cola Company, http://www.thecoca-colacompany.com/ourcompany/bios/bio_76.html.

9. Herman Vantrappen and Petter Kilefors, "Grooming CEO Talent at the Truly Global Firm of the Future," *Prism* 2 (2009): 91–105.

10. Partnership for a New American Economy, "The 'New American' Fortune 500," June 2011, http://www.renewoureconomy.org/sites/all/ themes/pnae/img/new-american-fortune-500-june-2011.pdf.

11. Joann Muller, "The Impatient Mr. Ghosn," *Single Articles* (blog), http://www.singlearticles.com/the-impatient-mr-ghosn-a779.html; David Ibison and James Mackintosh, "The Boss among Bosses," *Financial Times*, July 7, 2006, http://www.ft.com/cms/s/1/530603e4-0de3-11db-a385 -0000779e2340.html#axzz27tXNmlIA.

12. Carla Power, "India's Leading Export: CEOs," *Time*, August 1, 2011, http://www.time.com/time/magazine/article/0,9171,2084441,00.html.

13. R. Jagannathan, "5 Reasons Why Indian CEO's Are Making It Big on Global Stage," *FirstPost Business*, November 14, 2011, http://www .firstpost.com/business/5-reasons-why-indian-ceos-are-making-it-big-on -global-stage-130524.html.

14. Partnership for a New American Economy, "The 'New American' Fortune 500."

15. Lu Hong and Scott E. Page, "Groups of Diverse Problem Solvers Can Outperform Groups of High-Ability Problem Solvers," *Proceedings of the National Academy of Sciences* 101 (2004): 16385–89.

16. Irving L. Janis, *Victims of Groupthink: A Psychological Study of Foreign-Policy Decisions and Fiascoes* (Boston: Houghton Mifflin, 1972), p. 277.

17. Hong and Page, "Groups," p. 16389.

18. Jerome M. O'Connor, "Churchill's ULTRA Secret of the Century," September 2000, http://www.historyarticles.com/bletchley_park.htm.

19. Steve Lohr, "Netflix Awards $1 Million Prize and Starts a New Con- test," *Bits* (blog), *New York Times*, September 21, 2009, http://bits.blogs .nytimes.com/2009/09/21/netflix-awards-1-million-prize-and-starts-a

-new-contest/; "Academy Interview: Five Questions for Dr. Scott Page," NASA, June 30, 2010, http://www.nasa.gov/offices/oce/appel/ask-academy /issues/volume3/AA_3-6_F_interview.html.

20. Page and Hong, "Groups," p. 16386.

21. Warren E. Watson, Kamalesh Kumar, and Larry K. Michaelson, "Cultural Diversity's Impact on Interaction Process and Performance: Comparing Homogeneous and Diverse Task Groups," *Academy of Management Journal* 36 (1993): 590.

22. Katherine W. Phillips, Katie A. Liljenquist, and Margaret A. Neale, "Is the Pain Worth the Gain? The Advantages and Liabilities of Agreeing with Socially Distinct Newcomers," *Personality and Social Psychology Bulletin* 35 (2009): 336.

23. Joop de Jong, "The Dutch Golden Age and Globalization: History and Heritage, Legacies and Contestations," *Macalester International* 27 (2011), http://digitalcommons.macalester.edu/macintl/vol27/iss1/.

24. "Authoritarian Democracy in Singapore," BBC, http://www.bbc .co.uk/learningzone/clips/authoritarian-democracy-in-singapore/10102 .html.

25. Christopher Ong, "Albert Winsemius," Singapore Infopedia, http:// infopedia.nl.sg/articles/SIP_1457_2009-02-11.html.

26. Philip Bobbitt, *The Shield of Achilles: War, Peace, and the Course of History* (New York: Random House, 2002).

27. Ryan J. Maher, "A Priest Walks into Qatar and . . . ," *Washington Post*, July 20, 2008, http://www.washingtonpost.com/wp-dyn/content/ article/2008/07/18/AR2008071802558.html.

28. Thomas P. M. Barnett, *The Pentagon's New Map: War and Peace in the Twenty-First Century* (New York: Penguin, 2004).

29. Evgeny Morozov, "The Naked and the TED," *New Republic*, August 2, 2012, http://www.tnr.com/article/books-and-arts/magazine/105703/ the-naked-and-the-ted-khanna#.

30. Abraham Heschel, *The Prophets* (New York: HarperCollins, 1969), p. xv.

31. Pirkei Avot, 2:20.

INDEX

Page numbers in *italics* refer to illustrations.